101 KEY IDEAS

Astronomy

101 KEY IDEAS

Astronomy

Jim Breithaupt

TEACH YOURSELF BOOKS

For UK orders: please contact Bookpoint Ltd, 78 Milton Park, Abingdon, Oxon OX14 4TD. Telephone: (44) 01235 400414, Fax: (44) 01235 400454. Lines are open from 9.00–6.00, Monday to Saturday, with a 24 hour message answering service. Email address: orders@bookpoint.co.uk

For USA & Canada orders: please contact NTC/Contemporary Publishing, 4255 West Touhy Avenue, Lincolnwood, Illinois 60646–1975, USA. Telephone: (847) 679 5500, Fax: (847) 679 2494.

Long renowned as the authoritative source for self-guided learning – with more than 30 million copies sold worldwide – the *Teach Yourself* series includes over 200 titles in the fields of languages, crafts, hobbies, business and education.

British Library Cataloguing in Publication Data
A catalogue record for this title is available from The British Library.

Library of Congress Catalog Card Number: On file

First published in UK 2000 by Hodder Headline Plc, 338 Euston Road, London, NW1 3BH.

First published in US 2000 by NTC/Contemporary Publishing, 4255 West Touhy Avenue, Lincolnwood (Chicago), Illinois 60646–1975 USA.

The 'Teach Yourself' name and logo are registered trade marks of Hodder & Stoughton Ltd.

Copyright © 2000 Jim Breithaupt

Cover design and illustration by Mike Stones.

Typeset by Transet Limited, Coventry, England.
Printed in Great Britain for Hodder & Stoughton Educational, a division of Hodder Headline Plc, 338 Euston Road, London NW1 3BH by Cox & Wyman Ltd, Reading, Berkshire.

Impression number   10 9 8 7 6 5 4 3 2 1
Year   2006 2005 2004 2003 2002 2001 2000

Contents

Introduction

Welcome to the **Teach Yourself 101 Key Ideas** series. We hope that you will find both this book and others in the series to be useful, interesting and informative. The purpose of the series is to provide an introduction to a wide range of subjects, in a way that is entertaining and easy to absorb.

Each book contains 101 short accounts of key ideas or terms which are regarded as central to that subject. The accounts are presented in alphabetical order for ease of reference. All of the books in the series are written in order to be meaningful whether or not you have previous knowledge of the subject. They will be useful to you whether you are a general reader, are on a pre-university course, or have just started at university.

We have designed the series to be a combination of a text book and a dictionary. We felt that many text books are too long for easy reference, while the entries in dictionaries are often too short to provide sufficient detail. The **Teach Yourself 101 Key Ideas** series gives the best of both worlds! Here are books that you do not have to read cover to cover, or in any set order. Dip into them when you need to know the meaning of a term, and you will find a short, but comprehensive account which will be of real help with those essays and assignments. The terms are described in a straightforward way with a careful selection of academic words thrown in for good measure!

So if you need a quick and inexpensive introduction to a subject, **Teach Yourself 101 Key Ideas** is for you. And incidentally, if you have any suggestions about this book or the series, do let us know. It would be great to hear from you.

Best wishes with your studies!

Paul Oliver
Series Editor

Preface

This book is intended for those without a background in astronomy to read and learn about key ideas in astronomy. Astronomy is a subject with a long history and an ever-expanding future. The modern ideas of astronomy such as black holes, gravitational lenses, pulsars and quasars catch our imagination in a way that few other branches of science do, perhaps because astronomy is for people of all ages. Even the words and terms used, such as 'The Big Bang', convey the excitement of the subject and its capacity to inspire us. However, don't be deluded into thinking that astronomy long ago was any less exciting than it is now. The constellations themselves were mapped out centuries ago and given imaginative names to match the patterns seen. For many centuries, astronomers have tried to understand the universe and our place in it. The trials of Galileo started our present Scientific Age and present-day scientists are uncovering an astonishing picture of the origin of the universe. Long ago, events such as eclipses and the appearance of comets were very important in guiding the thoughts and actions of rulers in many countries. The astounding discoveries in modern science and astronomy enable us to place ourselves and our tiny planet in perspective.

Astronomy is a vast subject and this book provides a concise account of key ideas in an accessible and readable format. The 101 key ideas in this book are presented alphabetically, with diagrams where appropriate and cross-references where relevant. The key ideas cover the big ideas of modern astronomy such as the Big Bang and also the essential ideas and facts needed for beginners who want to learn about the night sky.

Antimatter

Matter consists of particles and antimatter consists of antiparticles. An antiparticle and a particle are produced from a photon of high energy radiation which ceases to exist as a result. An antiparticle has a rest mass equal and opposite to the rest mass of its particle counterpart and it has an equal and opposite charge to its particle counterpart if its particle counterpart is charged.

The first antiparticle to be discovered was the positron which is the antiparticle of the electron. An antiparticle such as the antiproton can be created (along with another proton) by making two protons collide at speeds approaching the speed of light. Antiparticles may combine with each other to form composite antiparticles such as antihydrogen atoms which each consist of an antiproton and a positron.

For a high energy photon to produce a particle and its antiparticle, the photon energy (hf) must be greater than or equal to the total rest energy of the particle and antiparticle (which is equal to $2m_0c^2$ where m_0 is the rest mass of the particle). When a particle and its corresponding antiparticle collide and annihilate each other, two photons of total momentum and total energy equal to the initial momentum and energy of the particle and antiparticle are created. In other words, whenever a particle is created, a corresponding antiparticle is created and whenever a particle is annihilated, a corresponding antiparticle is annihilated.

Galaxies consist of matter not antimatter. No observational evidence has been found in support of antimatter galaxies. Astronomers believe that the universe was created in a massive explosion known as the **Big Bang**, about 12 billion years ago. The energy from the Big Bang is thought to have created particles and antiparticles. Many more particles than antiparticles must have been created from radiation as the universe cooled after the Big Bang. This asymmetrical creation of more particles than antiparticles soon after the Big Bang ensured that all the antiparticles were annihilated by particles to form photons.

see also...

Big Bang; Dark Matter

Asteroids

The asteroids consist of minor planets and other bodies too small to be seen in orbit about the sun, mostly between Mars and Jupiter in a region known as the asteroid belt. The first asteroid, Ceres, was discovered in 1801 by Guiseppe Piazzi. Thousands of asteroids have since been discovered and their orbits measured. Some asteroids move along highly elliptical orbits with perihelions inside Mercury's orbit. Asteroids have been discovered on orbits which cross the Earth's orbit; the nearest recorded approach of such an asteroid to Earth in recent decades being in March 1989 at a distance of 700,000 km (0.005 AU). An asteroid is thought to have collided with the Earth about 65 million years ago, ending the Dinsosaur Age as a result.

Many asteroids move along orbits which are inclined to the Earth's orbit at much larger angles than any of the planets. Ceres, the largest asteroid, is about 770 km in diameter. Vesta and Pallas are the only other asteroids larger than about 500 km. Many asteroids have a maximum length of less than 10 km.

The mean distance of the asteroids from the Sun is about 2.8 AU. Two groups of asteroids known as the Trojans are on the same orbit as Jupiter, one group ahead by about 60° and the other group behind by about the same amount. Asteroids on orbits sufficiently elliptical cross the orbits of the inner planets, including Icarus which approaches the sun closer than Mercury. Asteroids far beyond Jupiter have also been discovered and may be part of the **Kuiper belt** of bodies beyond Neptune which is thought to be responsible for short-period comets.

The asteroids are thought to be composed of materials such as silica, iron and igneous rock. Images of asteroids were obtained from the Galileo spacecraft when it passed through the asteroid belt. These images revealed that asteroids are cratered but do not have a characteristic shape although the minor planets are thought to be almost spherical. None of the asteroids has an atmosphere as their gravity is too weak to retain gases released at the surface.

---

see also...

Planet; Planetary Orbits

Atmosphere (Earth's)

The Earth's atmosphere is composed of 78 per cent nitrogen, 21 per cent oxygen and less than 1 per cent argon, carbon dioxide and water vapour. The density of the atmosphere decreases with height from a mean value of approximately 1 kg m^{-3} at sea level to less than 10^{-9} kg m^{-3} at a height of 160 km. Above about 500 km, the atmosphere is non-existent. Over 50 per cent of the mass of the atmosphere lies below a height of 6 km.

The Earth's atmosphere protects living objects on Earth from the effects of bombardment by meteors which burn up when they enter the atmosphere; and from the effects of solar radiation consisting of particles, ultraviolet radiation and X-rays. Ultra violet radiation from the sun is filtered out by the ozone layer which is at a height of about 25 km. Charged particles from the sun are usually deflected away from the Earth by the Earth's magnetic field.

The Earth's atmosphere is transparent to electromagnetic radiation in two bands referred to as the visible window which covers the visible spectrum, and the radio window which covers frequencies from approximately 30 MHz to 100 GHz. Optical and radio telescopes at ground level can therefore detect light or radio waves from space unlike infra-red, ultraviolet and X-ray detectors which need to be on satellites above the atmosphere to detect radiation from space.

The molecules of the Earth's atmosphere scatter sunlight which is why the sky in daylight is bright in all directions. The sky is blue because the amount of scattering is greater, the shorter the wavelength of the light.

The ionosphere contains ions and electrons in a region from about 100 km to 300 km in height. Cosmic and solar radiation causes ionization of the atoms and molecules of the atmosphere in this region. Radio waves at frequencies below 30 MHz are reflected by the ionosphere because the presence of ions and electrons makes the ionosphere reflect radio waves like a metal plate does.

see also...

Meteors; Electromagnetic Radiation; Radio Telescope

Big Bang

The Big Bang theory of the universe is that the universe was created in a massive explosion from a point when space, time and matter was created. This event is thought to have occurred about 12 billion years ago. As the universe expanded, galaxies formed and moved away from each other. The universe continued to expand and still continues to expand. The distant galaxies are known to be rushing away from each other at speeds approaching the speed of light.

The Big Bang theory originated from the discovery by the American astronomer Edwin Hubble that the distant galaxies are receding from us at speeds in proportion to their distances. This observational relationship is known as Hubble's Law which states that, for a receding galaxy at distance d, its speed of recession $v = Hd$, where H is a constant of proportionality known as the Hubble constant. However, although Hubble's law is explained by the idea that the universe is expanding, the Big Bang theory was not accepted until the discovery of cosmic background microwave radiation in 1965. Before this discovery, many astronomers favoured the Steady State theory of the universe which supposed that the expansion of the universe is due to matter created in space between the galaxies which move away from each other as a result. The Steady State theory was discarded as it cannot explain the presence of cosmic microwave radiation from all directions in space.

Hubble deduced the link between speed of recession and distance as a result of measuring the red shift of more than 24 galaxies up to about 6 million light years away. Since then, many more galaxies have been measured for speed of recession and distance, giving results which continue to support Hubble's law and provide a more accurate value of H. The accepted value of the Hubble constant is now reckoned to be about 20 km s^{-1} per million light years. An increased value of H at greater distances would indicate the expansion of the universe is accelerating.

see also...

Expansion of the Universe; Hubble's Law

Binary Stars

Binary stars are stars that orbit each other. In most binary systems, two stars orbit about their common centre of mass due to their mutual gravitational attraction. Some binary systems consist of more than two stars; for example, Mizar and Alcor are two stars close to each other in the handle of the Big Dipper or Plough that can just be detected separately by an observer with sharp eyes without a telescope or binoculars. In fact, telescopic observations reveal that Mizar and Alcor are themselves binaries.

Binary stars were first discovered over two centuries ago when careful observations of certain double stars (which are stars that appear close together) showed them to consist of stars that orbit each other. For example, the brightest star in the sky, Sirius A, is accompanied by Sirius B which is much fainter. The two stars orbit each other with a time period of almost 50 years.

There are three main types of binary stars. *Visual binaries* are binary stars such as Sirius A and B that can be seen as separate stars directly or using a telescope. The plane of orbit is not necessarily perpendicular to the line of sight. *Eclipsing binaries* are binary stars that eclipse each other periodically because their orbits are edge-on to us. An eclipsing binary such as Algol, in the constellation Perseus, dips in brightness each time one of its two stars is eclipsed by the other. *Spectroscopic binaries* are binary stars that are detected as binaries only because the spectrum of light from the binary system is repeatedly shifted back and forth as the two stars approach and move away from us. The spectrum of light from a star comprises a continuous spread of the colours of the rainbow. Vertical lines cross the spectrum in certain positions. These lines are due to the absorption of light from the star's interior by the gases surrounding the star. The position of these lines is doppler-shifted to the red part of the spectrum when the star emitting the light is moving away from us, and to the blue part of the spectrum when the star is moving towards us.

see also...

Kepler's Laws of Planetary Motion; Spectra; Star Masses

Black Holes

Nothing can escape from a black hole, not even light. A black hole is a perfect absorber of all types of electromagnetic radiation (or any other form of radiation) just as a black surface is a perfect absorber of visible light. The idea of a black hole was first thought up by John Michell in 1783 although the term 'black hole' is of much later origin and was first coined by the American physicist, John Wheeler. Although Michell's general idea was correct, there was no evidence at that time that gravity affects light. In 1916 Albert Einstein predicted in his *General Theory of Relativity* that a strong gravitational field distorts space and time and bends light. Einstein calculated that light grazing the sun from a star was deflected by a thousandth of half a degree due to the sun's gravity. This prediction was confirmed by Sir Arthur Eddington in 1919 who led an expedition to South America to test the prediction by photographing stars that appeared close to the sun during a total solar eclipse.

The modern theory of black holes was started by Karl Schwarzschild who used Einstein's theory to prove that an object with a sufficiently strong gravitational field would prevent light from escaping. Schwarzschild showed that such an object is surrounded by an **event horizon**, a spherical envelope surrounding the object which nothing inside can pass through. Any object falling through the event horizon would disappear forever, leaving a fading image on the event horizon. The radius of the event horizon is known as the Schwarzcshild radius. The Schwarzschild radius of a black hole of mass $M = 2\ GM/c^2$, where G is the constant of gravitation from Newton's theory of gravity and c is the speed of light. The Earth would need to be compressed to a diameter of less than 18 mm to become a black hole.

Evidence for black holes has been obtained by astronomers. The central region of the galaxy M87 is rotating so fast that there is thought to be a massive black hole at its centre. The X-ray source, Cygnus XI, is a binary system consisting of a supergiant star accompanied by a very dense invisible star which may be a black hole pulling matter off its companion.

see also...

Einstein, Albert; Antimatter

Celestial Sphere 1 – Celestial Equator

Long ago astronomers imagined all the stars to be attached to an invisible sphere, referred to as the Celestial Sphere, surrounding the Earth, as illustrated below. The Earth spins about an axis through its poles at a constant rate of once every 24 hours, corresponding to 15 degrees per hour since each complete turn takes the Earth through 360 degrees. Ancient astronomers believed the Earth to be fixed at the centre of the Celestial Sphere which they imagined to be spinning at a constant rate of once every 24 hours, carrying the stars across the sky at a steady rate.

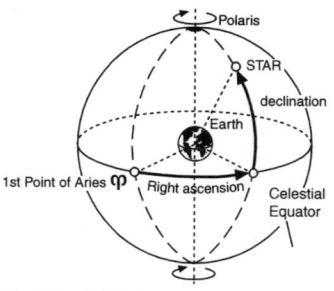

The Celestial Sphere

The Pole Star, also known as Polaris, is always directly above the Earth's North Pole. The axis of rotation of the Celestial Sphere may be imagined to pass through Polaris. The Celestial Equator is the projection of the Earth's equator onto the Celestial Sphere. A great circle on the Celestial Sphere is a circle that passes through both poles.

The position of a star on the Celestial Sphere is defined by two coordinates. The first is its *declination*, which is the angle to the star from the Celestial Equator along the great circle through the star. Stars that are north of the Celestial Equator are assigned positive values of declination; stars south of the Celestial Equator are assigned negative values. The second is its *right ascension*, which is the angle from a certain point along the Celestial Equator known as the First Point of Aries (γ) to the great circle passing through the star.

Right ascension is usually stated in hours corresponding to the time interval between the First Point of Aries crossing the observer's meridian (the great circle from north to south through the point directly above the observer and the Pole Star) and the star crossing the meridian from east to west.

see also...

Sidereal Time

Celestial Sphere 2 – Ecliptic

The Earth's axis is tilted towards the Pole Star. The Earth's North Pole is tilted towards the sun in June and away from the sun in December.

If the sun was much fainter, its path through the constellations as viewed from the Earth could be charted. This path is called the **ecliptic**. The Celestial Equator is at an angle of 23.5 degrees to the ecliptic because of the tilt of the Earth's axis.

At mid-summer in the northern hemisphere, the sun reaches its highest point on the ecliptic north of the Celestial Equator. This occurs when the sun lies in the constellation of Taurus, close to Gemini.

At mid-autumn in the northern hemisphere, the sun has moved round the ecliptic from its mid-summer position by 90 degrees. At this time of the year, it passes from north to south across the Celestial Equator in the constellation of Virgo.

At mid-winter in the northern hemisphere, the sun reaches its highest point on the ecliptic south of the Celestial Equator. At this time of

the year, the sun lies in the constellation of Sagittarius. At midday in mid-winter in the northern hemisphere, the sun is due south at its lowest point in the sky.

At mid-spring in the northern hemisphere, the sun passes across the Celestial Equator from south to north in the constellation of Pisces. This point of the year is known as the vernal equinox. The point where the ecliptic crosses the Celestial Equator is known as the **First Point of Aries** (γ).

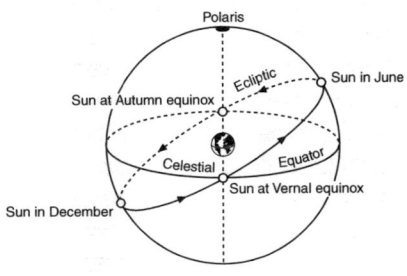

The sun on the Celestial Sphere

'Equinox' means equal day and night. At mid-spring and mid-autumn, darkness and daylight on each of these days are of equal duration.

> ### see also...
> *Celestial Sphere 1*

Celestial Sphere 3 – Circumpolar Stars

The Pole Star can be seen on any clear night in the northern hemisphere at any time of the year. The altitude of the Pole Star (i.e. its angle above the horizon) is the latitude of the observer. For example, an observer at the Earth's North Pole would see the Pole Star directly overhead. Stars near the Pole Star can also be seen on any clear night. The Earth's spinning motion makes each of these stars appear to move round on a circle centred on the Pole Star. A long-exposure photograph of such stars shows arcs centred on the Pole Star, each arc traced out by the image of a star on the film because the film moves relative to the star's image as the Earth turns. The altitude of such a star changes as it moves round the Pole Star. If the star never dips below the horizon, the star is referred to as **circumpolar**. For an observer at latitude L, a star that just dips to the horizon must be at an angle equal to L from the Pole Star. Any star further than this from the Pole Star cannot be circumpolar at this latitude. An observer at latitude L in the southern hemisphere would also be able to see an equivalent set of circumpolar stars.

Stars that are not circumpolar rise and set once every 24 hours. For example, in early winter in the northern hemisphere, the Orion constellation is seen on a clear night just after it has risen above the eastern horizon. By early morning, before sunrise, the same constellation is seen above the western horizon just before it sets. All stars that are not circumpolar rise in the east and set in the west because the Earth spins eastwards about an axis through its poles.

A star is said to **culminate** when it is at its highest point above the horizon. This occurs when the star passes from east to west across the observer's meridian which is the great circle on the Celestial Sphere from north to south through the Pole Star and the point directly above the observer. Every star culminates about four minutes earlier than the time of its culmination the previous night. This is because the Earth turns at a rate of one degree every four minutes and the orbital motion of the Earth around the sun is about one degree per 24 hours.

see also...

Celestial Sphere 1

Cepheid Variables

Cepheid variables are stars that vary smoothly in brightness between a lower limit and an upper limit with a constant period between 1 and 50 days approximately. This type of star was discovered by John Goodricke in 1784 who observed that the brightness of δ-Cephei in the constellation of Cepheus varied smoothly with a period of 5.4 days by about 1 magnitude. Other stars with brightness variations over similar periods were observed in different parts of the sky – all referred to as cepheid variables. The change of brightness of a cepheid variable is due to changes within the star that cause its diameter and hence its brightness to vary.

Henrietta Leavitt measured the variation of brightness of 25 cepheid variables in the Small Magellan Cloud which is a small irregular galaxy near the edge of the Milky Way galaxy. By plotting each star on a graph of mean apparent magnitude against period, she discovered that the mean apparent magnitude increased steadily with the period. Although the distance to the Magellan Cloud was not known by Leavitt, she did know that all the stars in the cloud were the same distance approximately from the sun (in the same way that everyone in New York is approximately the same distance from anyone in England). Hence, Leavitt concluded that the mean absolute magnitude of a cepheid variable increases with its period.

A year later, Enjar Hertzsprung was able to use the parallax method to measure the distance to a much closer cepheid variable and thereby was able to calculate the difference between the mean absolute and apparent magnitude for a cepheid variable, a difference that could then be applied to any cepheid variable to enable its absolute magnitude and hence its distance to be calculated. Cepheid variables were used by astronomers after Leavitt's discovery to determine the distances to stars far beyond the range of parallax measurements, including stars in other galaxies that could be seen individually.

see also...

Distance Measurement 2; Magnitude; Variable Stars

Clusters of Stars

Clusters of stars are groups of stars bound together by the force of gravity between the stars.

An *open cluster* ranges in size from a few light years to 50 or more light years in diameter. In general, blue stars predominate in open clusters and, because such stars have a much shorter life than red stars, open clusters contain relatively young stars. Prominent star clusters include M45, the Pleiades in Taurus, the Hyades in Taurus, and M44 Praesepe in Perseus. The Pleiades consists of blue stars surrounded by diffuse silvery clouds of dust over a region more than 20 light years in diameter. In contrast, the Hyades is an open cluster consisting of a large number of stars spread over a distance of more than 80 light years and moving along paths parallel to each other. Praesape, also known as The Beehive, contains about 200 stars in a region of diameter about 40 light years. More than a thousand open clusters have been observed in the plane of the Milky Way galaxy's spiral arms. In general, the stars in an open cluster will eventually move away from each other and the cluster will cease to exist.

A *globular cluster* is a tight spherical cluster of millions of stars held together by their own gravity. The diameter of a globular cluster is in the range from about 50 to 300 light years. Globular clusters in the Milky Way galaxy lie above and below the plane of the galaxy, distributed in all directions from the galactic centre. About 100 globular clusters have been observed in the Milky Way galaxy. Metal-poor red stars, referred to as population I stars, predominate in globular clusters, indicating that such clusters are very old. The gravitational attraction of all the stars in a globular cluster is sufficiently strong to prevent the stars from dispersing which is why globular clusters are so stable. The brighest globular cluster is Omega Centauri, a fourth magnitude object in the southern hemisphere containing a million or so stars in a spherical region of diameter about 160 light years at a distance from Earth of more than 20,000 light years.

see also...

Distance Measurement 2; Magnitude; Variable Stars

Comets

A comet is an object which moves around the sun in a highly elliptical orbit. A comet is prevented from leaving the solar system by the sun's gravity which slows it down as it moves away from the sun and speeds it up as it approaches the sun. When a comet is far from the sun, it is frozen, dark and invisible from the Earth. As it approaches the sun, its surface is heated by solar radiation, making it release glowing gas and dust which may each form a visible tail pointing away from the sun.

Long-period comets take hundreds or thousands of years to move round the sun. In 1997, Comet Hale Bopp returned to the inner solar system after a journey lasting several thousand years, taking it to a vast distance beyond Pluto. At its brightest, Comet Hale Bopp was easily visible in the night sky, including just before sunset and just after sunrise. Long-period comets are thought to originate in the **Oort cloud**, a collection of objects in orbit around the sun at a distance of about one light year. A nearby star could attract an object out of the Oort cloud into an elliptical orbit about the sun, causing the object to become a long-period comet.

Short-period comets take no more than a hundred years or so to complete one orbit. Halley's Comet takes about 76 years to go around the sun once. It was observed in detail by the unmanned space probe Giotto when it returned to the inner solar system in 1985. The observations caused astronomers to abandon the 'dirty snowball' model of a comet in favour of the 'chocolate peanut' model, as it was discovered that its surface is smooth and dark and was punctured from the interior by jets of gas which burst through its dark smooth surface as a result of internal pressure caused by solar heating.

Short-period comets may have been long-period comets which were attracted into short-period orbits by Jupiter on passing through the inner solar system. However, they may originate from a belt of thousands of rocks in the same plane as the ecliptic, the **Kuiper belt**, thought to orbit the sun not far beyond Pluto.

see also...

Asteroids; Planetary Orbits

Constellations 1 – Introduction

The constellations which we use to map the sky are patterns of stars defined by astronomers thousands of years ago in Ancient Greece. Other ancient civilizations also drew up maps of the sky in the form of constellations but it is the 88 constellations from Ancient Greece that are used now. Two stars that appear close to each other in the night sky may be further from each other than the nearer one is to the Earth. The apparent closeness is because they are almost in the same direction but they are unlikely to be at the same distance from Earth unless they are binary stars.

Except for constellations which are made up of circumpolar stars, the constellations visible in the night sky change during the year. This is because they are in the opposite direction to the sun when we view them from the night side of the Earth. The Earth orbits the sun once each year. As our home planet moves around the sun, our view of the night sky changes as the opposite direction to the sun changes. For example, the Orion constellation is a glorious winter constellation in the northern hemisphere because it is in the opposite direction to the sun in winter. There is no point in looking for Orion in summer as it is in the direction of the sun. This is because the Earth has moved around its orbit by about 180 degrees from its winter position. You can work out which constellations can be seen in each season using a star chart.

The belt of constellations which the ecliptic passes through is called the Zodiac. If the sun was much less bright and we could see the stars and the sun at the same time, the sun would be seen to move annually through the constellations as the Earth orbits the sun. The constellations which the sun would be seen to pass through are the constellations of the Zodiac. Because the planets orbit the sun in approximately the same plane, the planets observed from Earth are never far from the ecliptic and they therefore move through the constellations of the Zodiac.

see also...

Celestial Sphere 1

Constellations 2 – Prominent Features

The two observational exercises below are provided as 'starters' to help you find your way about the night sky in the northern hemisphere

1 Look north and locate the seven stars known as the Plough or the Big Dipper in Ursa Major. Using binoculars or a low-power telescope, the penultimate star of the 'handle' can be seen as two very close stars, Alcor and Mizar. At the other end of the Plough, the two stars Dubhe and Merak point to Polaris in Ursa Minor.

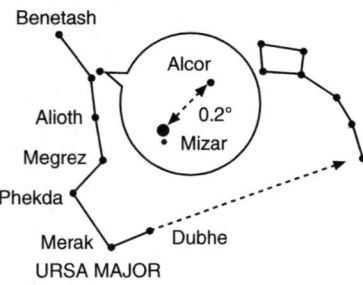

Benetash
Alcor
Alioth
0.2°
Megrez
Mizar
Phekda
Merak
Dubhe
URSA MAJOR

Finding the Pole Star

Continue your line of sight from the pointers of the Plough to Polaris and beyond to find Cassiopeia, easily located as it appears like a giant W in the night sky, at approximately the same angle from Polaris as the Plough. Just beyond Cassiopeia lies Andromeda which includes M31, the Andromeda galaxy, the only object outside the Milky Way that can be seen by the unaided eye.

2 Orion dominates the night sky in winter, rising in late autumn on the eastern horizon at about midnight and setting unseen at about midday. The top left-hand star of Orion is the red supergiant Betelgeuse. The star at its right foot is a blue-white supergiant, Rigel. Beyond the star at the left foot of Orion lies Sirius, the brightest star in the sky. From Orion: go westwards and higher to find the Pleaides in Taurus.

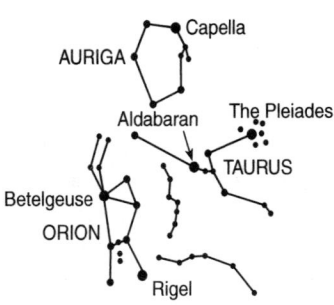

Capella
AURIGA
Aldabaran
The Pleiades
TAURUS
Betelgeuse
ORION
Rigel

Orion and its neighbours

see also...

Celestial Sphere 3; Supernova

Copernicus

opernicus re-established the heliocentric model of the universe which had been promoted by Aristarchus almost 2000 years earlier but rejected a century later in favour of Ptolemy's geocentric model. Ptolemy devised his model to explain the paths of the planets through the constellations. According to Ptolemy, the sun revolved around the Earth and the planets moved around on circles or 'epicycles' with centres that moved around the Earth which was fixed at the centre of the Celestial Sphere. This model predicted the paths of the planets through the constellations and it found favour with the Church as it fitted in with the idea that the human race has a special position in the universe.

Copernicus was born in Poland in 1472 and he developed a wide range of scholarly interests. The Ptolemaic model by then had become very elaborate. By the fifteenth century, no less than 80 spheres were needed. Copernicus devoted many years of his life to research on the issue of why alternatives to Ptolemy's model had been rejected. He reconstructed Ptolemy's model and eventually discovered it could be simplified to little more than a set of concentric circles, centred on the sun, representing the orbits of the planets. He feared that his revolutionary ideas would be ridiculed by his contemporaries and did not publish his work until shortly before he died. His work made little impact for many years because he presented it as a mathematical solution rather than a new scientific theory. However, in 1600 Giordano Bruno was burned at the stake by the established Church for promoting the Copernican model to support his view that the physical universe had no celestial boundary and that Heaven was not located beyond the Celestial Sphere. Bruno thus drew attention to the Copernican model 60 years after Copernicus died in 1543, resulting in Galileo's discovery of astronomical evidence in support of the Copernican model and its eventual acceptance by the Catholic Church in 1822, long after its acceptance by the scientific community.

see also...

Ptolemy's Planetary Model; Galileo

Cosmic Microwave Background Radiation

Microwave radiation is electromagnetic radiation in the wavelength range from about 1 to 100 mm. Microwave background radiation from all directions in space was discovered by Arno Penzias and Robert Wilson in 1965 when they were testing an aerial system designed to detect 74 mm wavelength radio signals from a satellite. They detected background radiation from all directions which led to further investigations and to the discovery that this background radiation has the same distribution of energy with wavelength as the radiation from an object at a background temperature of 2.7 K.

Cosmic microwave background radiation was released by matter in the early universe shortly after the Big Bang. Before the discovery of this microwave background radiation, the Big Bang theory was no more than a possible explanation of the expansion of the universe. Another prominent theory, known as the Steady State theory considered the universe to be infinite and eternal. In this theory, matter is created in the voids between galaxies, pushing them apart and forcing the Universe to expand as new galaxies formed. The existence of microwave background radiation can be explained by the Big Bang theory but not by the Steady State theory.

Background microwave radiation consists of photons released after the Big Bang when the Universe became transparent as it expanded and became cooler. Before this time, photons were continually being absorbed and re-emitted by atoms throughout the early universe. As the universe expanded and cooled, a crucial phase was reached when photons and atoms 'decoupled' from each other. This is thought to have occurred when the universe was little more than 100,000 years old and about a thousandth of its present size. The photons released at that stage and now being detected have stretched in wavelength as a result of the expansion of the universe as they travelled through space, so that they are now in the microwave range of the electromagnetic spectrum.

> *see also...*
>
> *Big Bang; Electromagnetic Radiation*

Craters

A crater is a saucer-shaped depression in the surface of a planet, usually with a rim higher than the surface beyond the crater. The origin of a crater may be volcanic or due to a metorite impact. Craters on the Earth are gradually eroded by wind and rain. A crater discovered off the coast of Mexico is thought to have been due to a meteorite impact 65 million years ago which caused the demise of the dinosaurs.

Craters also abound on the surface of Mercury, largely unchanged since their formation as Mercury has no atmosphere and therefore no wind and rain to cause erosion. The Caloris Basin on Mercury, a large ring structure over 1200 km in diameter, is thought to be due to a meteorite impact.

Lunar craters vary in diameter, the largest on the moon's nearside having a diameter of 295 km. The rim of a large crater may be thousands of metres above the crater floor. In comparison, the Grand Canyon in Arizona is insignificant as it is not much longer than about 10 km and is about 1.5 km deep.

Lunar craters are thought to be due to meteorite impacts in the early solar system as a result of debris orbiting the sun after the formation of the planets. Bright and dark patches visible on the lunar surface were thought to be mountains and seas respectively before Galileo used a telescope to observe craters in the dark and bright areas. The fact that the dark areas are less heavily cratered than the light areas implies that most of the impact craters were created before lava from the lunar interior burst through the surface and covered the light areas to form the 'seas'. Colossal meteorite impacts are thought to have caused these lava outflows which solidified after covering vast areas of the lunar surface. A further feature of some lunar craters are the 'rays' that lie across the surface spreading out from the crater. These lines are formed permanently from surface material ejected by the meteorite impact and thrown in different directions.

see also...
Moon; Mercury

Dark Matter

One of the biggest mysteries in science at the start of the twenty-first century is the whereabouts of most of the matter in the universe. This hidden matter is known as dark matter, sometimes called 'missing mass'. Dark matter is invisible matter hidden in galaxies or between galaxies but is known to be present as it slows galaxies down. It constitutes at least 90 per cent of the mass of the universe, yet it cannot be detected directly as it is not hot enough to emit light and does not absorb light.

The total mass of the galaxy can be estimated from its rate of rotation. A star at the outer edge of a spiral galaxy keeps moving around the galactic centre because it experiences a force of gravitational attraction to the galactic centre in the same way as a planet moving around the sun. However, the further a planet's orbit is from the sun, the longer it takes to go around, unlike the stars in the arms of a spiral galaxy which mostly go around with the same time period, regardless of distance. To produce the result that the time period is independent of radius, by applying Newton's theory of gravity to the motion of an outer star around a spiral galaxy, it is necessary to assume the galaxy includes much more matter in its spiral arms than can be accounted for by the stars there.

Thus the problem of missing mass arises, because if all the mass of a galaxy was in the stars of the galaxy, the galaxy should be much brighter than it actually is. In comparison with a typical star, such as the sun, the luminosity-to-mass output of a typical galaxy is less than one-tenth of that of a typical star. Since the light from a galaxy is due entirely to its stars, at least 90 per cent of the mass of a typical galaxy must be outside its stars and therefore hidden.

The search to find dark matter directly is an active field of research at present. Dark matter could be due to sub-atomic particles called neutrinos, produced and radiated from stars in vast quantities due to nuclear fusion. However, the mass of the neutrino is not yet known.

see also...

Luminosity; Nuclear Fusion; Star Masses

Distance Measurement 1 – Parallax

Two adjacent stars of the same brightness could be at vastly different distances from Earth; one could be much brighter and much further away than the other one.

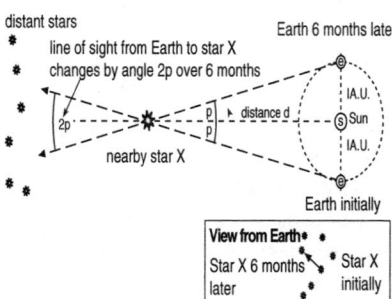

distant stars
line of sight from Earth to star X changes by angle 2p over 6 months
Earth 6 months later
2p
distance d
nearby star X
1 A.U.
S Sun
1 A.U.
Earth initially

View from Earth
Star X 6 months later
Star X initially

The Parallax method

The distances to stars less than a few hundred light years from Earth have been measured using the fact that such a star appears to shift its position annually against other stars in the same constellation. This effect is known as 'parallax' and it is due to the changing position of the Earth as it moves around its orbit. The line of sight from the Earth to the star changes direction as the Earth moves along its orbit, so the star's position moves against the background of stars in the same constellation,

shifting from one limit to another over six months, then returning over the next six months. The maximum shift of the line of sight to a nearby star is the angle between the two limits of the line of sight. This angle can be measured to within 0.02 arc seconds, where 1 arc second is one-sixtieth of one-sixtieth of one degree.

The parallax angle of the star is defined as half of the maximum shift. The distance to the star with a parallax angle of one arc second is defined as one parsec. This distance is equal to 3.26 light years. Because the parallax angle is equal to the angle between the lines from the sun and Earth to the star, it can be shown that the distance in parsecs to a star

$$= \frac{1}{\text{parallax angle in arc seconds}}$$

The parallax method cannot be used with ground-based telescopes for stars beyond about 100 parsecs observed from the ground as atmospheric refraction smudges stars out by about 0.01 arc seconds.

see also...

Magnitude; Hubble Space Telescope

Distance Measurement 2 – Beyond Parallax

The brightness of a star seen from Earth depends on its luminosity and its distance. The absolute magnitude of a star can be calculated from its apparent magnitude and its distance. The distances to sufficient stars within 100 parsecs were known to Ejnar Hertzsprung in 1911 and Henry Russell in 1913 independently for them to calculate the absolute magnitudes of these stars and plot them on a graph of absolute magnitude against temperature to create the Hertzsprung Russell diagram. Hertzsprung also made the first estimate of the distance to a cepheid variable which he then used to calibrate the magnitude v period relationship for cepheid variables, discovered in 1911 by Henrietta Leavitt. This important relationship has since been used to measure distances to other galaxies in which individual cepheid variables can be resolved. Thus, cepheid variables have been used as distance markers to other galaxies at distances of up to 1 million parsecs.

The distances to galaxies far beyond 1 million parsecs has been determined by measuring the red shift of each galaxy and then using Hubble's Law to calculate its distance. Hubble's Law that the red shift of a galaxy is proportional to its distance was discovered by Edwin Hubble in 1929 after he measured the red shifts of two dozen galaxies within 2 million parsecs of the Milky Way. He determined the distances to these galaxies which were beyond the range of the cepheid variable method by comparing their observed angular size and overall brightness with the average angular size and light output of the larger nearby galaxies at known distances.

The Hubble Space Telescope has been used to observe cepheid variables in galaxies at distances up to about 20 million parsecs, and these measurements have confirmed Hubble's Law as valid up to these distances. Further measurements using the Hubble Space Telescope to observe supernova in distant galaxies are being carried out to test Hubble's Law to much greater distances up to 1500 million parsecs.

see also...

Cepheid Variables; Hubble's Law; Magnitude; Red Shift

Earth

The Earth is the third planet out from the sun which is a middle-aged star in a large spiral galaxy we call the Milky Way. This galaxy contains millions of millions of stars. There are thought to be millions of millions of other galaxies. Planets have been detected in orbit around other stars in the Milky Way galaxy. The possibility that earth-like planets exist elsewhere in the universe seems likely. Life on Earth has evolved because our planet has water in liquid form and the surface is protected from ultraviolet radiation from the sun by the atmosphere. If the Earth was much nearer the sun, the oceans would evaporate; if the Earth was much further from the sun, the oceans would freeze. Life would probably not have evolved as it has in such circumstances. Fortunately, for much of the Earth's existence since its formation about 4500 million years ago, it has moved around the Sun on an almost circular orbit, keeping the same distance of 149.6 million km from the sun to within 0.01 per cent. This distance is defined as **one astronomical unit** (AU).

The Earth's shape is that of a slightly flattened sphere as its polar diameter is 6357 km which is about 13 km less than its equatorial diameter. The kilometre was originally defined as one ten-thousandth of the distance from the Equator to the North Pole. This definition has been replaced by another based on the speed of light which gives a distance from the equator to the North Pole as 9986 km. The Earth's interior has a solid high-density core about 2500 km in diameter surrounded by a liquid core of about 8000 km in diameter. Above the core, the solid mantle extends to the crust which is a solid layer of thickness about 40 kilometres deep. The core is thought to contain iron whereas the mantle consists of less dense silica-based materials. The Earth's magnetic field is generated in the liquid core, probably as a result of electrically charged matter driven by thermal convection. None of the other terrestial planets possesses a magnetic field which suggests that their interiors are solid now.

see also...

Planet; Atmosphere (Earth's)

Eclipses of the Moon

Eclipses of the sun or moon occur when the Earth, the sun and the moon are directly in line with each other. When the moon is in exactly the opposite direction to the sun, a lunar eclipse is seen because the moon passes through the Earth's shadow. The eclipse is total if the moon passes completely into the umbra of the Earth's shadow, otherwise only a partial lunar eclipse is seen. In a total lunar eclipse, the moon does not disappear completely as some sunlight is bent by the Earth's atmosphere into the umbra. Because this effect is greatest for red light, the lunar disc at totality may appear dull red or the colour of copper. Just before or after totality, beads of light known as 'Baily's beads' due to crater rims and similar features on the lunar surface may be seen at the edge of the lunar disc.

A lunar eclipse does not happen every full moon. The reason is that the moon's orbit is inclined at five degrees to the Earth's orbit around the Sun. At most times when a full moon occurs, the moon passes above or below the Earth's shadow. The two points where the moon's orbit passes through the

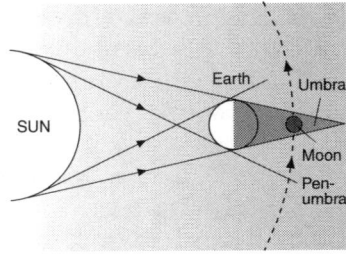

An eclipse of the moon

plane of the Earth's orbit are called 'nodes'. For a solar or lunar eclipse to occur, the moon must be at one of

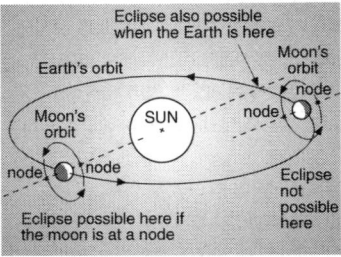

Nodes

these two nodes and the nodes must be on the line from the sun to the Earth.

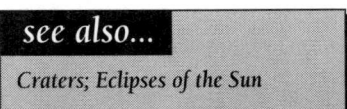

see also...

Craters; Eclipses of the Sun

Eclipses of the Sun

clipses of the sun or moon occur when the Earth, the sun and the moon are directly in line with each other. When the moon is directly between the sun and the Earth, a solar eclipse is seen on the Earth because the moon's shadow covers part of the Earth's surface. The region of total shadow, the umbra, extends from the moon like a cone, covering an area on the daylit-side of the Earth of about 300 km in diameter. Because the Earth spins on its orbit, this area of totality moves across the surface at a speed of the order of thousands of kilometres per hour. Anyone outdoors in the path of totality experiences no more than a few minutes of darkness. At totality, the solar corona consisting of hot gases around the sun's disc can be seen. These gases stretch millions of kilometres into space. The region of partial shadow, the penumbra, covers an area of about 5000 km on the Earth. Anyone observing the eclipse from this region would see a partial solar eclipse as the moon's disc gradually covers then uncovers some but not all of the sun's disc. An annular eclipse is seen when the moon is further than usual from the Earth so the umbra of its shadow does not reach the Earth. The sun is then seen as a ring or 'annulus' around the dark lunar disc. **Note that the sun should never be observed without the aid of a suitably recommended dark filter**.

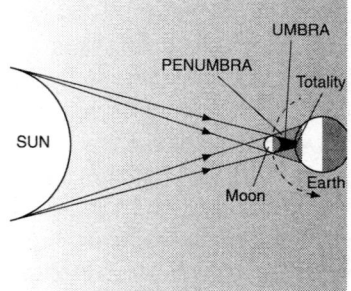

An eclipse of the sun

A solar eclipse does not happen every new moon which is when the moon passes between the Sun and the Earth. The reason is that the Moon's orbit is inclined at five degrees to the Earth's orbit around the Sun. At most times when a new moon occurs, the Moon's shadow passes above or below the Earth.

see also...

Eclipses of the Moon; Solar Activity

Einstein, Albert

Albert Einstein started a revolution in physics in 1905 when he published the *Special Theory of Relativity* in which he proved that energy and mass are interchangeable in accordance with the equation $E = mc^2$, where c is the speed of light in free space. He also showed that no object can travel faster than light. In the same year, he also published a paper which proved the existence of atoms and another paper in which he proved that light consists of photons which are packets of electromagnetic waves, each consisting of energy in proportion to the frequency of the waves. Einstein was in his mid-twenties at the time and was employed as a full-time patent officer in Berne. He became a full-time university lecturer in 1909 and moved to Berlin in 1913 as director of a research institution specially set up for him.

In 1916 Einstein published the *General Theory of Relativity* in which he predicted the existence of black holes and the bending of light by gravity. His theory was successfully tested by Sir Arthur Eddington who photographed stars near the sun during the total eclipse of 1919. Eddington discovered that the position of the stars in line with the edge of the sun's disc were slightly displaced as Einstein had predicted, and that the amount of displacement was the same as Einstein had calculated it should be. The successful test of Einstein's General Theory meant that concepts such as absolute space and absolute time were no longer correct. Space and time are inter-linked and affected by gravity. Following a conference of leading scientists in London to discuss the implications of Einstein's theory, Einstein became an international celebrity when the conference was reported the next day in *The Times* newspaper. Einstein was awarded the Nobel Prize for Physics in 1921 for his 1905 paper on the nature of light. He emigrated to America in 1933 and died in 1955. The General Theory of Relativity has had important consequences for astronomy and cosmology, including the discovery of black holes, gravitational lensing and the Big Bang theory of the origin of the universe.

see also...

Big Bang; Black Holes; Gravitational Lensing

Electromagnetic Radiation

Electromagnetic radiation consists of electromagnetic waves which are perpendicular electric and magnetic waves in phase with each other. Electromagnetic waves are self-propagating as the electric wave generates the magnetic wave which generates the electric wave. The spectrum of electromagnetic radiation is detailed below. All electromagnetic waves travel through a vacuum at the same speed as light does, namely 300,000 km per second.

Electromagnetic radiation is quantized which means it consists of packets of electromagnetic waves. Each packet is a photon of electromagnetic radiation. The energy E of a photon is given by the equation $E = hf$, where f is the frequency of the waves and h is a constant known as the Planck constant.

	Wavelength range	Space sources	Detection of space sources
Radio	> 0.1 m	Sun, interstellar gas clouds, radio galaxies, quasars	Ground-based radio telescopes
Microwaves	0.1–1 mm	Sun, cosmic background	Ground-based microwave detectors
Infra-red	1mm–700 nm	Sun, cool stars, instellar dust clouds	Satellite-based infra-red detectors
Visible	400–700 nm	Sun, stars, galaxies, interstellar gas clouds	Ground-based telescopes, eye, CCD camera
Ultraviolet	1–400 nm	Sun, stars, interstellar gas clouds	Satellite-based UV detectors
X-rays	< 1 nm	Sun, X-ray binaries, supernovae, intergalactic gas	Satellite-based X-ray detectors
Gamma rays	< 1 nm	Sun, interstellar gas, gamma ray bursts	Satellite-based gamma ray detectors

see also...

Infra-red; Astronomy; Radio Astronomy; Ultraviolet Astronomy

Escape Speed

A rocket needs to attain a speed of about 11 km/s to escape from the Earth's surface and reach the moon or beyond. This minimum speed is known as the **escape speed** from the Earth's surface. If the engines of a rocket are not powerful enough, the rocket does not attain escape speed and its kinetic energy is then too small to enable it to overcome the pull of gravity due to the Earth so it falls back to Earth.

The escape speed of an object from a point in a gravitational field is defined as the minimum speed an object needs to have to escape from that point in the field to infinity.

It can be shown that the escape speed from a point at distance r from the centre of a planet is equal to $\sqrt{(2gr)}$, where g is the gravitational field strength at that point.

At the surface of the Earth g = 9.80 N kg^{-1} and r = 6370 km approximately. Hence the escape speed is $\sqrt{(2 \times 9.80 \times 6370 \times 1000)}$ = 11,200 metres per second.

At the surface of the moon, g = 1.62 N kg^{-1} and r = 1740 km, hence the escape speed from the moon = 2380 m/s. The much smaller escape speed from the moon is the reason why the *Apollo* astronauts who set foot on the moon were able to return from the lunar surface to their lunar orbiters in modules much smaller than the giant Saturn rockets which were needed to escape from Earth.

The Earth has an atmosphere whereas the moon does not. This is because gas molecules in the Earth's atmosphere move at speeds far below the escape speed of 11.2 km/s and hence are unable to escape from the pull of the Earth's gravity. Gas molecules released on the moon would have a similar range of speeds as molecules in the Earth's atmosphere as the temperature range on the moon is similar to that on the Earth. However, the gas molecules would escape from the moon because the escape speed is much lower from the lunar surface.

see also...
Strength of Gravity

Evolution of Stars

The evolution of a star is the sequence of stages it passes through from its formation as a protostar to the end of its life as a light-emitting object. A star forms from dust and hydrogen gas clouds as a result of the inward pull of gravity of matter in the clouds on other matter in the clouds. As the matter of the protostar becomes denser, gravitational energy is converted to thermal energy and the temperature of the star increases until the protostar is hot enough for nuclear fusion to commence. High energy radiation released at this stage heats the protostar even more and it becomes a main sequence star.

hydrogen nuclei being fused into helium nuclei in its core. Radiation released in this process exerts pressure on the layers of the star outside the core. The gravitational pull on each layer of a star is balanced by the outward force due to radiation pressure. When all the hydrogen nuclei in its core has been used, the core collapses and the outer layers swell out and cool to become a red giant. At this stage, the helium nuclei in the core fuse, forming nuclei as heavy as iron. When this process ends, the 'entire' star collapses and heats up to form a 'white dwarf' if its mass is less than 1.4 solar masses. Planetary nebulae seen around some stars is thought to be glowing matter thrown off as a result of this collapse to a white dwarf. If the mass of a star exceeds 1.4 solar masses, known as the 'Chandresakhar limit', the star collapses completely and then explodes as a supernova.

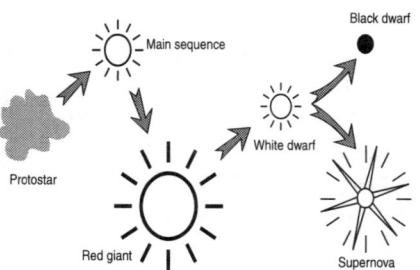

The evolution of a star

A star stays in the same position on the main sequence for much of its life, emitting radiation as a result of

see also...

Hertzsprung Russell Diagram; Nova; Nuclear Fusion; Red Giant; Supernova; White Dwarf

Expansion of the Universe

Very strong scientific evidence supports the theory that the universe is expanding as a result of a massive explosion about 12 billion years ago which created space and time. As the universe expanded, galaxies formed and moved away from each other as the universe continued to expand and continues to expand. The distant galaxies are known to be rushing away from each other at speeds approaching the speed of light.

The discovery that the universe is expanding was made in 1929 by the American astronomer Edwin Hubble. He proved from his observations that the distant galaxies are receding from us at speeds in proportion to their distances. This statement is known as Hubble's Law and may be stated as the following equation: For a receding galaxy at distance d, its speed of recession $v = Hd$, where H is known as the Hubble constant.

Since 1929, the speed of recession and distance for many more galaxies has been measured, giving results which continue to support Hubble's law and provide a more accurate value of H. The accepted value of the Hubble constant is now reckoned to be about 20 km s^{-1} per million light years. Hubble's Law is explained by assuming that the universe is expanding so the further away a galaxy is, the faster it is receding from us. The expansion of the universe can be explained by the Big Bang Theory and another theory, known as the Steady State Theory, which supposes that matter is being continuously created between the galaxies, pushing them apart in the process. However, the discovery of cosmic background microwave radiation from all directions in space could be explained only as radiation created in the Big Bang, so the Steady State Theory was discarded.

It is not yet known if the expansion of the universe will slow down and reverse or if it will continue to expand for ever. The mean density of the universe needs to be determined for if it exceeds a value known as the 'critical density', reversal is inevitable.

see also...

Big Bang; Cosmic Microwave Background Radiation; Hubble's Law

Eyepieces

The eyepiece of a telescope is designed to channel light entering the telescope from a distant object into the eye of the observer and also to enable the observer to see a magnified view of the image formed by the objective if the object is not a point object.

The magnifying power of a telescope is equal to the ratio of the focal length of the objective to the focal length of the eyepiece. The shorter the focal length of an eyepiece, the greater its power is said to be as it makes the magnifying power of the telescope greater. By using eyepieces of different focal lengths, the magnifying power of a telescope can be altered. This is useful when observing the Moon or a planet as these are not point objects so they appear larger through a telescope. As the field of view is reduced if the magnifying power is increased, a low-power eyepiece is used for general observations where a wide field of view is desirable. When an object of interest such as a planet is located, the eyepiece can then be replaced by a high power eyepiece to give a more detailed view.

The width of the eyepiece lens normally exceeds about 8 mm, which is the width of the eye pupil in darkness. Light from a distant object which enters the telescope and passes through the eyepiece thus enters the eye. An eyepiece is usually a combination of two lenses at a separation less than or equal to the average of the two focal lengths. This arrangement eliminates chromatic aberration, the splitting of white light into colours, which would spoil the observed image. A high-quality eyepiece also eliminates spherical aberration which is distortion of an image due to the outer part of a lens focusing light differently from the central part.

A telescope fitted with a film camera allows long exposure images to be formed, enabling objects too faint to be seen directly to be observed. The eyepiece position is adjusted to enable a real image to be formed on the camera film. A charge coupled device (CCD) camera replaces an eyepiece and a film camera, with the pixel array of the CCD camera located where the light from the objective is brought to a focus.

see also...

Magnification; Telescopes

Galaxies 1 – Classification

A galaxy is an assembly of millions of millions of stars, held together by their mutual gravitational attraction. Galaxies may be classified broadly according to appearance as either spiral, elliptical or irregular.

A spiral galaxy has spiral arms that wind around its centre. The Milky Way galaxy is a spiral galaxy with a diameter of the order of 100,000 light years. Spiral galaxies vary in size from about one-tenth of the size of the Milky Way galaxy to not much larger than the Milky Way. Blue stars predominate in the spiral arms whereas red stars predominate in the central regions. An elliptical galaxy is egg-shaped without any spiral arms. Elliptical galaxies vary in size from dwarf ellipticals which are about one-fiftieth of the size of the Milky Way galaxy to giant ellipticals which are about five times as large as the Milky Way galaxy. Irregular galaxies do not have a characteristic shape.

Edwin Hubble made a very detailed study of galaxies using the 250 cm reflector telescope in California in the 1920s. He devised the so-called 'tuning fork' diagram shown below in which he classified elliptical galaxies on a scale E0 (spherical) to E7 (cigar shaped) and spiral galaxies according to whether or not the centre was bar-shaped and according to the tightness of the spiral arms on a three-point scale a, b and c.

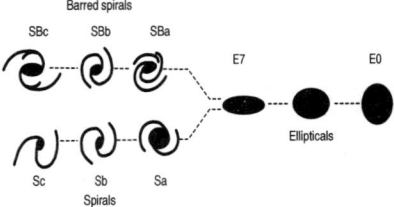

Hubble's tuning fork diagram

Astronomers now reckon that many elliptical galaxies may have formed as a result of mergers between spiral galaxies causing the destruction of the spiral arms. The discovery in 1994 of a very distant giant elliptical galaxy containing significant quantities of dust suggested some elliptical galaxies contain new stars and were not formed from spiral galaxies merging.

see also...

Stars 4

Galaxies 2 – The Local Group

Galaxies range in size from dwarf galaxies much smaller than the Milky Way to giant galaxies considerably larger than the Milky Way galaxy.

The sun is one of a million million stars in the Milky Way galaxy which is over 100,000 light years in diameter. The Milky Way galaxy is one of a group of neighbouring galaxies called the **local group**. The largest member of the local group is the Andromeda galaxy (M31), a spiral galaxy about 2 million light years away. The Milky Way galaxy is also a spiral galaxy. The sun is in one of the spiral arms of the Milky Way galaxy. The nearest star to the Sun is just a few light years away.

The universe is thought to contain millions of millions of galaxies, each containing millions of millions of stars. Deep space photographs reveal that galaxies group together in clusters, with each cluster containing thousands of galaxies, and that clusters link together to form superclusters of galaxies separated by vast empty regions or voids. The Milky Way galaxy and Andromeda and other galaxies in the local group form a cluster about 3 million light years

across. The other galaxies in the local group include the Triangulum Spiral M33 and several irregular galaxies including the Magellan Clouds. The Triangulum galaxy M33 is also referred to as the Pinwheel galaxy because of its resemblance to a snapshot of a pinwheel firework. It lies about ten degrees from M31 in the small constellation of Triangulum, named because the three most prominent stars of the constellation form a small skinny triangle. The Fornax galaxy in the constellation of Fornax lies much closer than M31 but it can only be seen with a large telescope because it is much smaller and fainter than M31.

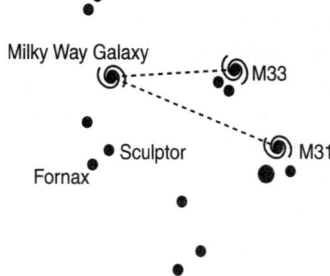

The local group of galaxies

see also...

Galaxies

Galaxies 3 – Clusters and Superclusters

Most galaxies belong to a cluster. The next nearest cluster of galaxies to the local group is the Virgo cluster which contains over 3000 galaxies in a patch of the sky no more than about ten degrees wide. The Virgo cluster is over 10 million parsecs away, around 20 times further than the Andromeda galaxy. Many more clusters of galaxies have been observed at distances over 6000 million light years away.

By counting the total number of galaxies above a certain brightness, Edwin Hubble estimated the total number at about 3000 million million. Even with as many as 1 million galaxies in every cluster, the number of clusters would be in excess of 3000 million — a number more than the entire human population of the world.

Clusters are distributed in all directions. A three-dimensional model of the distribution of all clusters at known distances reveals the presence of superclusters which are clusters of clusters and of voids which are large empty regions. In addition, clusters arranged in filaments and in sheets have been discovered. The so-called 'Great Wall' is a sheet of galaxies about 200 million light years away. Another very large concentration of clusters known as the 'Great Attractor' is thought to be attracting us and the Virgo cluster. However, on a larger scale, little evidence has been found for groupings and structures and the distribution of clusters smooths out. Making measurements over distances of the order of 100 million light years, the number of galaxies is the same in different directions. The distribution of galaxies can be likened to the distribution of material in a sponge. The holes represent voids and the sponge material represents galaxies. The distribution of sponge material is very uneven on a 'hole size' scale but it is uniform on a much larger scale. In 1999, an Anglo-German team of astronomers confirmed this picture after mapping the universe in infra-red radiation up to 300 million light years away.

> ## see also...
> *Galaxies 1 and 2*

Galileo

alileo was born in 1564 in Pisa, Italy. As the son of a nobleman, Galileo was educated in a monastery and in 1595 became Professor of Mathematics at the University of Padua, one of Europe's leading universities at that time, in what was then the Republic of Venice. His Venetian paymasters allowed to him to follow his own interests and his discoveries on motion would have been sufficient to win long-lasting recognition. In 1609, reports reached him about the invention of an optical device, the telescope, for making distant objects appear larger and closer. Within a short time, he had designed and constructed his own telescopes. He used his telescopes to study heavenly bodies and observed ten times as many stars as can be seen directly without a telescope. He found that the surface of the moon is heavily cratered and he discovered the four innermost moons of Jupiter.

Galileo's astronomical discoveries were widely reported in Europe. Galileo hoped his observations and conclusions in support of the Copernican model would be accepted by the Church but in 1613 he was reprimanded by the Church for his views. In 1623, a new pope was elected so Galileo travelled to Rome to try to persuade him to withdraw the ban on the Copernican system, imposed in 1616. The new pope refused to accept any challenge to the existing order and so Galileo decided to set out his views and his support for the Copernican model in print in Italian, for all to read. He completed his work, *Dialogue on the Two Chief World Systems* and published it in Florence in 1632. Galileo's book was an instant best-seller and the Church reacted rapidly by banning it.

Galileo was summoned to appear before the Holy Office of the Inquisition in Rome on 12 April, 1633. The judges decided that Galileo had broken the 1616 ban and had acted deceitfully. He was forced to recant and made to spend the rest of his life under house arrest at his home in Florence where he died in 1642. The publicity that Galileo's trial attracted perhaps helped to promote the Copernican system.

see also...

Copernicus; Ptolemy's Planetary Model; Telescopes

33

Gamma Ray Bursts

I f your eyes could detect gamma rays, you would be surprised to observe occasional flares in the sky, each perhaps lasting as long as a minute or so. In the mid-1960s, the US Defense Department launched a series of satellites to spot any secret nuclear weapons tests in space by the USSR. Instead, the satellites unexpectedly discovered gamma ray bursts from random directions in space. The cause of these gamma flashes remained a source of speculation among astronomers until 1991 when a satellite carrying a gamma ray observatory was launched into space from the space shuttle Atlantis. The observatory detected bursts at a rate of about one per day from random directions. The randomness of the locations of the gamma ray bursts means that the phenomenon must be associated with the universe at large, not with any particular part of the universe.

A new satellite, Beppo-SAX, was launched in 1996 which included a wide-angle X-ray telescope and a gamma detector on board. In February 1997, the detector was triggered by a gamma ray burst, designated GRB 970 228, that lasted

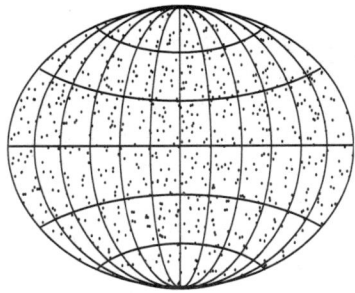

Sky map of gamma ray bursts

over a minute from the direction of Orion. The X-ray telescope was used to pinpoint the source before it became too faint to detect. The result was confirmed within 12 hours by astronomers using the 4.2 m William Herschel Telescope at La Palma in the Canary Isles. Beppo-SAX located another gamma burst, GRB 970 508, in May 1997 which was found to have a red-shift of 0.84. At last the question was settled about whether or not gamma ray bursts are near or far, as a red-shift of 0.8 clearly corresponds to a distance of the order of billions of light years. The cause of gamma ray bursts is not yet known.

see also...

Electromagnetic Radiation; Red Shift; Supernova

Gravitational Lensing

light travels in a straight path –
unless it passes through a
strong gravitational field. Albert
Einstein worked out that gravity bends
a light beam. He had already worked
out that gravity cannot be
distinguished from accelerated
motion. The essential idea about the
effect of gravity on light is not too
complicated if we consider a light
beam passing through opposite-facing
portholes of an accelerating rocket.

Light bending in an accelerating rocket

If the path of the beam could be made
visible to an observer in the rocket,
the observer would see a curved path.
The acceleration of the rocket causes
the observer to see a curved light
path. As acceleration cannot be
distinguished from gravity, it therefore
follows that gravity should cause a
light beam to curve too. Einstein

worked out that a light beam that
passes near the edge of the sun
should be deflected by an angle of
1.75 seconds of arc. Einstein's
prediction was tested successfully in
1918 by a team of astronomers led by
Arthur Eddington who observed and
measured the deflection of stars
occulted by the solar disc during a
total solar eclipse.

Can gravity cause distorted images of
objects in space? In 1979, a double
quasar Q0957 and 561 was
discovered. Because the signal from
each part fluctuated in the same way,
it was realized that the two parts
were in fact two images of a single
quasar. The quasar is hiding behind
this very large mass but we can see
two images of the quasar because its
light, skimming opposite edges of the
intervening galaxies, is bent round.
More recently, the Hubble Space
Telescope has revealed images of
faint galaxies distorted and spread
out into streaks behind clusters
which act as enormous lenses.

see also...

Eclipses of the Sun; Einstein

Hertzsprung Russell Diagram

The Hertzsprung Russell (HR) diagram is a graph on which each star is plotted on the diagram according to its absolute magnitude and its temperature. The diagram was devised independently by the Danish astronomer Enjar Hertzsprung in 1911 and the US astronomer Henry Russell in 1913. Stars range in absolute magnitudes

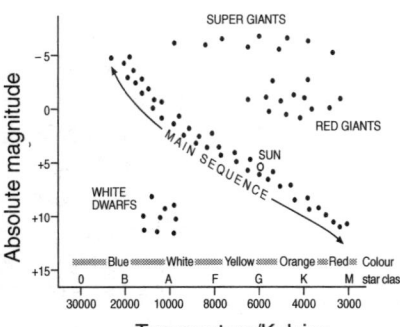

The Hertzsprung Russell diagram

from about +15 which is 10,000 times less powerful than the sun, to about -10 which is a million times more powerful than the sun. Most of the stars on the diagram lie on a diagonal belt which runs from the bottom right-hand corner to the top left-hand corner. This is referred to as the Main Sequence. The very powerful M-class stars, high above the Main Sequence,

are referred to as giants or supergiants. These stars are much larger than the sun which is why they are collectively referred to in this way. This can be deduced from the fact that an M-class star is cooler than the sun so it emits less light per unit area of its surface. It must, therefore, be much larger than the sun as it emits more power. Giant stars lie about 5 magnitudes above the Sun on the HR diagram. Supergiants lie about 5 magnitudes higher than giant stars.

The group of very hot faint stars below the Main Sequence are referred to as white dwarfs. The surface temperatures of these stars is much higher than that of the sun so a white dwarf emits much more light per unit area than the sun does. However, compared with the sun, a white dwarf is a feeble emitter of light so it must be much smaller in diameter than the sun. The HR diagram provides information about how stars evolve from birth to death, and how giants and dwarfs form part of this lifecycle.

see also...

Evolution of Stars; Magnitude; Red Giant; White Dwarf

Hubble Space Telescope

The Hubble Space Telescope (HST) is a reflector telescope which is in orbit above the Earth's atmosphere at a height of over 500 km. It is fitted with a charge coupled device (CCD) camera and other instruments which enable it to observe infra-red and ultraviolet images as well as visible images. The HST was launched into this orbit in 1990 from the Space Shuttle. Since its 2.4 m mirror was corrected in 1993, the HST has captured stunning images of many objects in space. These images are much much clearer and brighter than those produced from ground-based telescopes. Images seen through ground-based telescopes are smeared out due to atmospheric refraction as well as due to diffraction by the objective mirror. The HST is able to see much more detail than a ground-based telescope. In addition, it is unaffected by atmospheric absorption of light so it collects more light than a similar telescope on the ground. A major surprise for astronomers using the HST was the discovery of clusters of galaxies in a direction previously thought to be empty space.

Using the HST, astronomers have been able to measure the distances to stars and galaxies more accurately, repeating the parallax method for nearby stars then refining the link between the mean absolute magnitude of a cepheid variable and its time period. This more accurate link has then been used to measure with greater accuracy the distances to other galaxies by observing individual cepheid variables in such galaxies. One of the first experiments carried out using the HST involved observing individual cepheid variables in M100, a spiral galaxy with a known red shift which gave its speed of recession at 1400 km/s. The distance to M100 was thus measured at 55 million light years, giving a value of the Hubble constant of 25 km s^{-1} per million light years, accurate to 20 per cent. This gave a value for the age of the universe of about 8000 million years. Subsequent measurements give a value for the Hubble constant of 20 km s^{-1} per million light years, corresponding to an age of 12,000 million years.

see also...

Cepheid Variables; Clusters of Stars; Galaxies 1; Telescopes 1

Hubble's Law

Edwin Hubble used the 2.5 m reflector telescope on Mount Wilson in California to estimate the distances to two dozen galaxies of known red shifts within 2 million parsecs of the Milky Way galaxy (1 parsec = 3.26 light years). His results, published in 1929, showed that the red shift increased with distance. By plotting the results on a graph of red shift against distance, it was clear that the red shift and hence the speed of recession is in proportion to the distance. This relationship is known as Hubble's Law. The constant of proportionality in the relationship is known as the Hubble constant, H.

Speed of recession v = Hd, where d is the distance

Further measurements of more galaxies were made by Milton Humason using the 250 cm telescope at Mount Wilson. By 1935, Hubble and Humason were able to publish the results for more than 140 galaxies out to distances of more than 300 million parsecs moving away at speeds over 40,000 km s^{-1}. The results confirmed Hubble's findings of 1929 that the red shift increased with distance. Hubble and Humason estimated the Hubble constant to be 160 km s^{-1} per million light years. Subsequent measurements using bigger telescopes and improved detectors have reduced the Hubble constant to a present-day value of about 20 km s^{-1} per million light years.

Hubble's law is an experimental law valid for a limited range of measurements. Its possible explanations caused great controversy for half a century after it was announced. It is now accepted that Hubble's Law follows because the universe is expanding from a primordial explosion between 10,000 and 15,000 million years ago. This explosion, known as the Big Bang, was the origin of space and time. The value of H has a very profound significance as it is used to estimate the age of the universe.

see also...

Big Bang; Expansion of the Universe; Red Shift

Infra-red Astronomy

nfra-red radiation is electromagnetic radiation of wavelength in the range from about 700 nm to about 1 mm.

Infra-red radiation from objects in space is absorbed by water vapour in the atmosphere so infra-red telescopes are located either at high altitude where the humidity is low or on satellites above the atmosphere. An infra-red telescope contains a large concave reflector which focuses the radiation onto an infra-red detector. The telescope must be cooled to stop infra-red radiation being emitted by the telescope itself. The three metre Infra-red Telescope facility in Hawaii is sited there because the location is very dry.

Infra-red radiation is emitted by objects in space that are not hot enough to emit light. In addition, dust clouds in space emit infra-red radiation. Thus, infra-red telescopes are able to provide images of objects and dust clouds in space that cannot be seen using optical telescopes. The infra-red astronomical satellite (IRAS) was in orbit for ten months in 1983 during which time its 60 cm reflector provided images of dust clouds

around nearby stars and found that distant galaxies emit considerable amounts of infra-red radiation.

The Infra-red Space Observatory (ISO) provided infra-red images of objects and dust clouds in space for over two years after its launch in 1995. A 0.85 m infra-red telescope, the Space Infrared Telescope facility, is to be launched into orbit in 2002.

see also...

Atmosphere; Electromagnetic Radiation

Jupiter 1 – The Planet

Jupiter is the largest planet in the solar system. Its mass is 318 times that of the Earth and its diameter is a little more than 11 times the Earth's diameter. It orbits the sun at a mean distance of 5.2 AU once every 11.9 years, varying in distance from 5.0 to 5.4 AU. Jupiter is at opposition every 13 months when the Earth catches up and overtakes it. Viewed from Earth through a telescope, it is seen as a slightly flattened disc crossed by coloured light and dark parallel belts. Long-lived spots are seen in its belts, including the Great Red Spot which was recorded almost two centuries ago. Jupiter is the shape it is because it is a spinning ball of fluid, rotating once every ten hours. Its rapid rotation causes its equatorial diameter to be significantly larger than its polar diameter. The belts are due to the combined effects of its rapid rotation and thermal convection from its interior. The spots are thought to be long-lived whirlpools.

Analysis of the spectrum of light from Jupiter has enabled astronomers to deduce the presence of ammonia, methane and hydrogen in its atmosphere. Jupiter is thought to have a solid iron/silica core surrounded by hydrogen under such enormous pressure that it behaves as a metal. Hydrogen in molecular form exists between this metallic hydrogen and the atmosphere. Jupiter's atmosphere is at a temperature of about −110°C which is more than expected on the basis of its distance from the sun. Thus, Jupiter is reckoned to be releasing thermal energy from its interior, probably generated as a result of gravitational energy released when it was formed or perhaps released relatively recently if it continues to contract. In July 1994, Comet Shoemaker-Levy crashed into Jupiter in multiple fragments, causing temporary dark patches in its atmosphere where the impacts took place. Jupiter has been studied extensively by space probes, notably the two Voyager missions which flew past Jupiter within a few months of each other in 1979, sending close-up pictures of the surface of Jupiter and of its moons and of a faint ring system not previously known about.

see also...

Comets; Planet; Planetary Orbits

Jupiter 2 – The Moons

Jupiter is known to have at least 16 moons at distances from the planet in the range of 130,000 km to over 20 million km. Galileo discovered the four largest moons, Io, Ganymede, Callisto and Europa, now known as the Galilean moons of Jupiter. These four moons can be seen using a telescope as spots of light in the equatorial plane of the planet. Their positions change from one night to the next because they orbit the planet in days rather than years. Voyager 2 sent back detailed pictures of the Galilean moons as it flew past. More detailed images have come from the spacecraft Galileo which went into orbit around Jupiter in 1995.

Io, the innermost galilean moon, is about the same diameter as the Earth's moon. Its orbital period is 1.8 days and its orbital radius is 0.42 million km. Volcanic activity on Io astonished astronomers when it was first observed on the pictures from Voyager 2. Io's molten interior and eruptions from its surface are thought to be caused by repeated stretching and squeezing on account of the combined effect of Jupiter's gravity and that of the outer moons. Io has a mean density about the same as the Earth's moon.

Europa, the second Galilean moon outwards from Jupiter, is smaller than Io and takes 3.6 days for one complete orbit. Its orbital radius is 0.67 million km. Its smooth surface is coloured with brown patches and covered in cracks. The mean density of Europa is about the same as that of the Earth's moon.

Ganymede, the next galilean moon from the centre, is the largest moon of Jupiter. It orbits Jupiter every 7.2 days. Its orbital radius is 1.1 million km. Faults and bright craters are on the surface which is thought to be a thick crust of ice. The mean density of Ganymede is 1.9 times that of water.

Callisto, the outermost galilean moon, is not quite as large as Ganymede and has a similar mean density. It orbits Jupiter once every 16.7 days and its orbital radius is 1.9 million km. Its heavily cratered surface is a crust of ice and rock.

> ### see also...
> *Craters; Galileo*

Kepler's Laws of Planetary Motion

Johann Kepler was an astronomer who lived in Prague in the first three decades of the seventeenth century. He measured the orbit of each planet and determined how long each planet took to make one complete orbit of the sun. From his measurements, he formulated three laws which describe the motion of the planets.

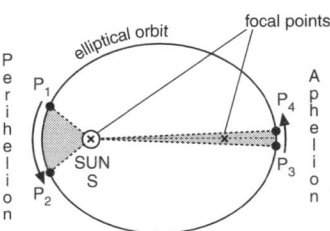

Time from $P_1 \rightarrow P_2$ = Time from P_3 to P_4
if area SP_1P_2 = area SP_3P_4

Kepler's Laws

The 1st Law states that each planet moves along an elliptical orbit in which the sun is at one of the two focal points of the ellipse.

The 2nd Law states that the rate of progress of an imaginary line from the sun to a planet varies as the inverse of the square of the distance from the planet to the sun.

Kepler knew that the perihelion distance of Mars is $0.9 \times$ the aphelion distance. He found the planet's apparent progress at aphelion was $0.81 \times$ its apparent progress at perihelion which he realized is the same as the square of the perihelion distance to the aphelion distance. See diagram.

The 3rd Law states that the cube of the mean radius of orbit is proportional to the square of the period for each planet. This law may be written as the equation below in which the period is in years and the mean radius is in astronomical units.

$$Period^2 = Mean\ radius^3$$

Kepler's laws were proved mathematically by Sir Isaac Newton using Newton's theory of gravity which brings the mass of the central body in to give the equation below, where the mass is expressed in solar masses.

$$Mass \times period^2 = Mean\ radius^3$$

see also...

Newton, Isaac; Planetary Orbits

Luminosity

The luminosity of a star is a measure of its light output, usually expressed either in watts or in terms of the luminosity of the sun which is 400 million million million million watts. Thus a star with a luminosity 100 times that of the sun emits light at a rate of 40,000 million million million watts.

In 1920, Sir Arthur Eddington gathered enough data on binary stars to show that the greater the mass of a star, the greater its light output. For main sequence stars, star masses vary from about 0.1 solar masses at the foot of the main sequence where stars are less than one-ten-thousandth as luminous as the sun to about 30 solar masses at the top where the stars are over 1 million times as luminous. Eddington showed that for main sequence stars, the luminosity is approximately proportional to the cube of the mass. In other words, a star that has a mass twice that of the sun emits approximately eight times ($= 2 \times 2 \times 2$) as much light as the sun; a star with a mass three times that of the sun emits about 27 ($= 3 \times 3 \times 3$) times as much light as the sun; and a star with a mass ten times that of the sun emits about 1000 times ($= 10 \times 10 \times 10$) as much light as the sun.

The absolute magnitude of a star is determined by its luminosity. Eddington's luminosity v mass relationship can be used to find the luminosity and hence the absolute magnitude of binary stars of known mass at unknown distances. Beyond about 200 parsecs, the parallax method of measuring distance is not possible as the parallax angle is too small. Binary stars beyond this distance can nevertheless be studied spectroscopically to enable their masses to be found. Knowing the mass of a binary star enables its luminosity and hence its absolute magnitude to be estimated if it is a main sequence star. Its distance can then be estimated using the relationship between absolute magnitude, distance and observed magnitude. In addition to enabling binary distances to be worked out beyond 200 parsecs, the luminosity v mass relationship provides an endpoint which has to be explained by any model of the processes and structure of a star.

see also...

Magnitude; Stars 4

Magnification

A telescope magnifies any distant object except point objects such as stars. The magnification of a telescope, also known as its magnifying power, is a measure of how much wider an object appears when viewed through the telescope in comparison to when it is viewed directly. For example, the angular width of the moon is about 0.5 degrees. A telescope with a magnification × 12 would make the moon appear to be six degrees wide.

The magnification of a telescope in normal use is equal to the ratio of the focal length of the objective to the focal length of the eyepiece. Thus a reflector telescope with a mirror of focal length 600 mm and an eyepiece of focal length 40 mm would have a magnification of × 15 (= 600 mm/ 40 mm). Changing the eyepiece for a different eyepiece of focal length 30 mm would therefore increase the magnification to × 20 (= 600 mm/ 30 mm).

The magnification of a telescope should exceed the ratio of the objective diameter to the diameter of the eye pupil, otherwise not all the light entering the telescope from a point object passes into the eye of the observer. Since the eye pupil diameter is, in dark conditions, approximately 8 mm, the magnification should be greater than or equal to one-eighth of the objective diameter in millimetres. If the magnification is too great, the image quality is poor because the eyepiece is too powerful for the telescope and it causes image distortion. As a general rule, the magnification should not exceed the numerical value of the objective diameter in millimetres. Thus a telescope with an objective diameter of 120 mm should have a magnification of at least × 15 and no more than × 120.

see also...

Eyepieces; Telescopes 2

Magnitude

Our present system of classifying stars according to brightness is thought to have originated with Hipparchus in the third century BC who divided stars into six categories according to brightness.

The brightest stars were referred to as stars of the 'first magnitude' and the faintest, those just visible to the unaided eye, were classified as 'sixth magnitude'. Astronomers in the nineteenth century measured the light intensity for stars of differing magnitudes and put the magnitude scale on a scientific footing by defining a brightness increase of 5 magnitudes as exactly 100 times as much light. Each brightness increase of 1 magnitude therefore corresponds to 2.512 times as much light, since 2.512 x 2.512 x 2.512 x 2.512 x 2.512 is equal to 100. The classification from first to sixth magnitude was extended at both ends of the scale so stars seen only with the aid of a telescope are assigned to the magnitude range beyond 6 and very bright stars are assigned to the magnitude scale beyond m =1 using the range from 1 to 0 and below 0.

To make a true comparison of the luminosity of different stars, it is necessary to calculate the magnitude each star would have if it was at the same distance from the solar system. This standard distance is chosen as ten parsecs for convenience. The magnitude a star would have at this distance is referred to as the **absolute magnitude M** of the star.

The absolute magnitude M of a star can be calculated from the magnitude m of the star if its distance is known. Note that m is also referred to as the 'apparent' magnitude. The calculation is based on the principle that the intensity of light at a certain distance from a point source varies with the inverse of the square of the distance. The inverse square law means that the light intensity changes by a factor $(d/10)^2$ on moving from distance d to 10 parsecs from the star. If Δm presents the corresponding magnitude difference, then $100^{\Delta m/5} = (d/10)^2$. Using base 10 logarithms therefore gives $\Delta m = 5 \log d - 5$ hence $M = m + 5 - 5 \log d$.

see also...

Luminosity

Mars 1 – Observations from Earth

Mars is the nearest planet to the Earth beyond Earth's orbit, orbiting the sun once every 1.88 years at a mean distance from the sun of 1.52 AU. Its diameter is just over half that of the Earth and its day is 37 minutes longer than a day on Earth. Its axis of rotation is tilted by about 25 degrees to its orbital axis, changing the direction in which it points very gradually. Its surface gravity is 0.38 times that of the Earth and the escape speed from its surface is 5.0 km/s which is less than half the escape speed from the Earth.

Mars is at opposition once every two years and 50 days. Its distinctive red colour makes it easy to pinpoint in the night sky as it moves gradually through the constellations of the ecliptic. Its retrograde motion for a month or so as the Earth catches up and overtakes it before and after opposition can be plotted easily against the star background. When observed through a telescope at or near opposition, Mars appears as a reddish disc with dark patches and white polar caps. The caps vary in extent according to the changing seasons on Mars which are due to the tilt of its rotation axis. Because its orbit is elliptical, its distance from the sun varies from 1.38 AU to 1.66 AU so its distance from Earth at opposition varies between 0.38 AU and 0.66 AU. Mars can be seen best from Earth when it is at opposition at its least distance from Earth. This most favourable opposition occurs every 14 or 15 years in August or September.

Mars has two moons, Phobos and Deimos, thought to be captured asteroids. Phobos orbits Mars once every 7 hours 39 minutes at a height of about 6000 km. Deimos orbits the planet at a height of about 20,000 km once every 30 hours.

see also...

Planet; Planetary Orbits

Mars 2 – Observations from Space Probes

Space probes to Mars, such as Mariner 9 in 1971 and the Mars Global Surveyor in 1997, have revealed the martian surface is strewn with rocks, is heavily cratered and has large deserts which cause global dust storms. The reddish colour of Mars, once thought to be due to its atmosphere, is now known to be due to the colour of the minerals on its surface. Mountains, canyons, volcanoes, valleys, ridges and dried river beds have been extensively photographed from orbiting space probes. Olympus Mons, a volcano 600 km wide at its base and 23 km high, can be seen from Earth through a sufficiently powerful telescope. The ice caps are thought to be solid carbon dioxide and water ice. The Northern ice cap recedes in spring, suggesting the loss of carbon dioxide ice into the atmosphere, leaving a cap of water ice. The dried river beds are clear evidence that water once flowed in large quantities on Mars.

The Martian atmosphere consists of carbon monoxide, carbon dioxide, oxygen and hydrogen at a pressure which is less than 1 per cent of the Earth's atmospheric pressure at sea level. Although at times thin clouds form in the atmosphere, heat radiation from the surface at night causes the surface temperature to fall from a maximum of 10°C at midday on the equator down to –75°C at night and as low as –120°C on the ice caps.

Liquid water is not in evidence on Mars as water ice turns directly into water vapour at very low pressures. However, the tilt of the axis of rotation may have been as much as 35 degrees in the past which would have caused much hotter conditions and higher atmospheric pressure on the planet. The Viking spacecraft that landed on Mars in 1976–7 found no evidence of life in the soil samples that were tested although it is possible that life in the form of microbes exists in sub-surface pockets inside craters. No positive evidence for life was found from further tests in 1997 by the robot rover vehicle which was carried to Mars by the Mars Pathfinder. However, rounded pebbles and rocks discovered by the rover means that running water was once present on Mars.

> ### see also...
> *Atmosphere (Earth's); Craters*

Mercury

Mercury is the nearest planet to the sun and is therefore very difficult to observe from Earth as it is never more than about 28 degrees from the sun which means that it sets no later than two hours after sunset and it rises no earlier than two hours before sunrise. Its mean distance from the sun is 0.39 AU and it orbits the sun once every 88 days. Its orbit is inclined at about seven degrees to the Earth's orbit and is highly elliptical, varying in distance to the sun from 0.31 to 0.47 AU.

Mercury's diameter is about 0.4 × the Earth's diameter and the surface temperature is thought to rise to about 350°C at midday and fall to about −170°C at night. Mercury rotates once every 59 days which is about two-thirds of its year. Its surface gravity at 0.36 times the Earth's surface gravity is not strong enough to enable it to retain an atmosphere. Its surface was photographed by Mariner 10 which flew past it twice in 1974 and again in 1975, sending back images of craters, valleys and mountains. The Caloris Basin, a ring with a diameter of 1300 km with a depressed floor bounded by mountains up to two kilometres high, is thought to have been caused by an impact due to a massive meteorite.

Transits of Mercury are seen periodically when Mercury passes directly between the Earth and the sun. An image of the sun projected on a suitable surface shows a black dot gradually moving across the solar disc during a solar transit. A solar transit does not occur each time Mercury passes between the Earth and sun because the inclination of Mercury's orbit to the Earth's orbit exceeds the angular width of the solar disc.

The perihelion of the orbit of Mercury gradually advances at a rate of 0.16 degrees per century. This effect, discovered in 1859, cannot be explained completely using Newton's theory of gravitation. In 1916, Einstein showed that the effect can be completely explained from his General Theory of Relativity.

see also...

Einstein, Albert; Planet; Planetary Orbits

Messier Objects

Charles Messier was an eighteenth-century French astronomer who discovered over 100 objects in the night sky which were neither comets, stars nor planets. These objects were referred to as 'nebulae' because they are fuzzy (unlike stars which are point objects), and do not move relative to the stars like planets and comets do. Messier discovered them in his searches for comets which he knew changed position among the stars. The fuzzy objects that did not change position among the stars were catalogued by Messier and are known as Messier objects. For example, the Crab Nebula, now thought to be the remnants of a star that exploded in the eleventh century, was the first object to be catalogued by Messier and is therefore known as M1. Messier catalogued the Andromeda galaxy as the Andromeda nebula M31 as he did not know about galaxies. Forty Messier objects are galaxies, each consisting of millions of millions of stars. The universe consists of countless galaxies receding from each other as a result of the Big Bang. The other Messier objects are either globular clusters in the halo of the Milky Way galaxy or open clusters or clouds of glowing gas and dust in the spiral arms of the Milky Way galaxy. Messier's catalogue of 110 nebulae was superseded in 1888 by the New General Catalogue (NGC) of Nebulae and Clusters of stars. However, the nebulae in Messier's catalogue are still referred to as Messier objects and the associated numerical designation still used. Some named Messier are listed below:

M1   The Crab Nebula – the remains of a supernova in Taurus about 6500 light years away.

M27   The Dumbbell Nebula – a planetary nebula less than 1000 light years away.

M31   The Andromeda galaxy; a spiral galaxy over 2 million light years away

M42   The Orion Nebula – glowing gas and dust clouds in Orion, less than 2000 light years away.

M45   The Pleiades – an open cluster in Taurus more than 400 light years away, consisting of more than 250 stars in a diffuse nebula.

M104   The Sombrero galaxy – a spiral galaxy in Virgo, about 40 million light years away.

see also...

Galaxies 3; Supernova

Meteors and Meteorites

A meteor is a particle from space that enters the Earth's atmosphere at high speed and burns up completely, appearing as a streak of light referred to as a 'shooting star'. Meteors vary in colour and duration although most meteors appear then disappear within a fraction of a second. A meteorite is a rock from space that hits the ground as it does not burn up completely in the atmosphere. Many particles are in orbit around the sun, ranging in size from several kilometres to less than a millimetre. Some of these particles are debris from comets which were fragmented on passing through the inner part of the solar system.

Meteors that enter the atmosphere from any direction are known as sporadic meteors. At certain times of the year, meteor showers occur when the Earth crosses the orbit of a comet or the remains of a comet. Seen from the ground, the tracks of the meteors of a meteor shower appear to radiate from a definite part of a constellation, referred to as the radiant of the meteor shower. This effect is because particles in the same orbit as the comet cause the shower. When they hit the Earth, they do so from a certain direction corresponding to the direction of the orbit as seen from Earth. Prominent meteor showers include the Leonids in November and the Perseids in late July. An annual meteor shower is particularly intense when the particles are bunched together along the orbit and the Earth passes through the bunch.

Most meteorites are either iron, stony material or a mixture of both. Their origin may be on account of collisions between larger objects in the Asteroid belt, causing chips of rock to fly off into earth-crossing orbits. The largest meteorite on record is a 60,000 kg object that hit the ground in south-west Africa. An impact by a very large meteorite is thought to have ended the Dinosaur Age millions of years ago. In 1969, a meteorite disintegrated in the skies above Allende in Mexico, scattering thousands of fragments over a wide area. Subsequent analysis of these fragments led to the theory that the meteorite was from a nearby supernova explosion billions of years ago.

see also...

Atmosphere (Earth's); Comets; Supernova

Milky Way Galaxy

The sun is just one of millions of millions of stars in the Milky Way galaxy, a spiral galaxy which is about 100,000 light years in diameter. The sun lies in one of the arms of the 'spiral' galaxy. The galaxy is rotating, taking about 240 million years for each complete rotation. The fact that the outer spiral arms rotate at much the same rate as the inner arms indicates the presence of dark matter in the galaxy.

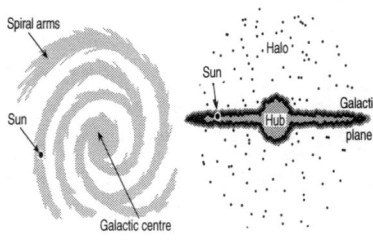

The sun in place

Globular clusters lie above and below the plane of the spiral arms in a halo. Dust clouds prevent light reaching us from the galactic centre. However, radio waves from the galactic centre and the spiral arms are unaffected by dust and have been used to map out the structure of the Milky Way galaxy. Hot blue metal-rich stars,

referred to as 'population I' stars, predominate in the spiral arms whereas metal-deficient red giants, referred to as 'population II' stars, predominate in the globular clusters and at the galactic centre. Population II stars are thought to have formed when the universe was little more than 1000 million years old. Short-lived massive population II stars, formed in the spiral arms in that era, have since exploded as supernovae, leaving metal-rich dust clouds from which the population I stars have formed.

On a clear night, the Milky Way appears to the unaided eye as a faint irregular diffuse band of light across the sky. The centre of the galaxy lies in the direction of the constellation of Sagittarius through two spiral arms which lie between the sun and the galactic hub. The spiral arm containing the sun is referred to as the Orion arm after the constellation Orion which lies in the same arm as the sun.

see also...

Dark Matter; Galaxies 3; Supernova

Models of the Universe

In 1929, Edwin Hubble discovered that the further away a galaxy is, the faster it is receding from us. This is explained by the theory that the universe is expanding. Over two centuries before Hubble's discovery, Sir Isaac Newton realized that if the universe is finite, then the stars could not be stationary otherwise they would all be attracted by each other's gravity into a great mass but Newton had no evidence for such motion. So Newton thought that the universe must be static and infinite.

Newton's model of the universe went unchallenged until a very simple problem was raised which became known as 'Olber's Paradox'. This is based on a very simple observation, namely that the night sky is dark not bright! This seemingly trivial observation was first analyzed by Heinrich Olbers in 1826. He proved mathematically that the sky would be permanently bright if the universe consisted of an infinite number of stars. He reasoned that the universe must be finite because the sky is dark. Since a finite static universe would collapse, Olbers reckoned it must be expanding.

Einstein used his general theory of relativity to predict that a finite static universe without boundaries is possible, like the Earth's surface except in four dimensions not two. Einstein had to introduce a new type of repulsive force which acted only over a cosmological scale into his equations. This repulsive force was thought necessary by Einstein to overcome the force of attraction due to gravity which would make a finite static universe collapse.

However, in 1927, Georges Lemaitre, a Belgian priest and mathematician, discovered 'expanding universe' solutions to Einstein's equations without the necessity of the cosmological force. Hubble's discovery in 1929 that the distant galaxies are receding meant that the cosmological force was not necessary. Lemaitre also discovered that his solutions had been worked out five years earlier by the Russian mathematician Alexander Friedmann who died in 1925. His most interesting solution is one in which the universe expands then contracts.

see also...

Einstein, Albert; Hubble's Law; Newton, Isaac

Moon 1 – Observations from Earth

The moon orbits the Earth at a mean distance of 384,000 km once every 27.3 days. Its appearance as seen from Earth passes through a cycle of phases once every 29.5 days. This occurs because the fraction of its sunlit surface visible from Earth changes as it moves around the Earth.

At new moon, it is seen as a slender crescent with the crescent tips pointing eastwards. The moon is then positioned between the Earth and the sun so almost all its sunlit surface faces away from the Earth. As the moon moves eastwards along its orbit, the fraction of its sunlit surface visible from Earth gradually increases (or 'waxes') until its entire sunlit surface is visible from Earth at full moon. From the Earth, it is then in the opposite direction to the sun. As its eastward motion along its orbit continues, the fraction of its sunlit surface visible from Earth decreases (or 'wanes') until it is seen as a slender crescent with tips pointing westwards before disappearing and reappearing as a new moon.

The moon's surface as seen from Earth has bright and dark patches, referred to

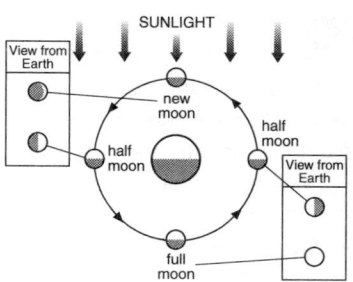

Phases of the moon

as 'mountains' and 'seas' (or maria) respectively. Craters can be seen all over the lunar surface, most easily observed at the edge of the sunlit part of the surface where the crater rims cast long shadows across the surface. The 'seas' are relatively smooth areas thought to be due to lava outflow from impacts of massive meteorites early in the moon's history when its interior was still partly molten. The hemisphere of the moon facing the Earth never changes because the moon rotation period is precisely the same as its orbital period. This effect is due to the 'tidal' gravitational forces exerted on the moon by the Earth which distort the moon slightly and make it turn.

see also...

Craters

Moon 2 – Libration

The moon always keeps the same face to the Earth because it spins on its axis at the same average rate as it turns on its orbit. Seen from the Earth, the moon moves eastwards through the constellations near the ecliptic, spinning counterclockwise slowly on its axis so it always presents the same face to the Earth. However, we see more than half its surface because, at different times, some of its surface 'around the sides' and 'over' the poles can be seen. These effects are known respectively as 'libration of latitude' and 'libration of longitude'.

Libration of latitude occurs because the distance from the moon to the Earth varies from 360,000 km to 406,000 km because the moon's orbit is elliptical, not circular. The speed of the moon on its orbit therefore varies slightly, being greatest when it is closest to the Earth and least when it is furthest. Because the rate at which it spins never varies whereas its rate of turning due to its orbital motion does vary, some of its far-side surface becomes visible from Earth.

When it is moving faster than average, because it turns on its orbit faster than it spins, we see some of its surface at the western edge of the lunar disc not usually visible. When it is moving slower than average, because it turns on its orbit slower than it spins, we see some of its surface at the eastern edge of the lunar disc not usually visible.

Libration of longitude occurs because its orbit is tilted at an angle of almost seven degrees to the Earth's orbit around the sun. When the moon is at the highest point of its orbit (i.e. furthest above the plane of the Earth's orbit), we can see under its South Pole and, therefore, some of its surface beyond this is visible from Earth. When the moon is at the lowest point of its orbit (i.e. furthest below the plane of the Earth's orbit), we can see over its North Pole and, therefore, some of its surface beyond this is visible from Earth.

see also...

Celestial Sphere 2

Moon 3 – Space Missions

The lunar surface is a hostile environment as there is no atmosphere to protect it from ultraviolet radiation or high energy particles from the sun. The temperature at the equator varies from about 130° C at midday to about −180° C at night. Objects on the surface near the equator are therefore subjected to much greater temperature changes than objects on the Earth. Less variation of temperature occurs in the polar regions as the sun is lower in the sky than at the equator. Water ice may be present in the depths of craters near the poles, perhaps capable of providing water for humans who might one day colonize the moon.

Space missions to the moon became possible only as a result of the successful construction of multi-stage rockets such as the powerful Saturn rockets which carried the Apollo astronauts into space. Space missions to the moon have included:

Unmanned Russian Luna probes: In 1959 Luna 2 crashed into the Mare Imbrium and Luna 3 went around the Moon and sent back the first photographs of its far side. In 1966 Luna 13 made the first successful soft landing on the Moon, and in 1970 Luna 16 brought back lunar rock samples.

Unmanned American Ranger probes: The Orbiter probes and the American Surveyor lander vehicles in the 1960s all provided detailed pictures of much of the lunar surface.

Apollo missions: Several missions in the late 1960s and early 1970s; most notably the historic Apollo 11 mission in 1969 which took Neil Armstrong (and Buzz Aldrin) to the moon to become the first person to set foot on the moon. Seismometers left on the moon from Apollo and other space missions have detected moonquakes which are much less powerful than earthquakes. In addition, emission of gas from below the surface in certain regions has been observed from Earth.

Unmanned post-Apollo missions: These have included Clementine in 1994 which detected polar ice at the moon's South Pole, and the Lunar Prospector in 1998 which confirmed the presence of large quantities of ice at the lunar poles.

see also...

Craters

Neptune

eptune was observed by Galileo who thought it was a star because its position against nearby stars did not seem to change. However, in 1843, John Adams in England and Urban le Verrier in France concluded independently that the unexplained speeding up then slowing down of Uranus must be due to an unknown outer planet pulling on Uranus. The position of the unknown planet, to be called Neptune, was worked out from the motion of Uranus and this prediction was confirmed by Johann Galle in 1846.

Neptune appears from Earth as a featureless blue disc, moving through the constellations at a rate of about two degrees per year, taking 164 years to orbit the sun at a mean distance of 30 AU. Its diameter is about the same as Uranus although its mean density is greater at 1.3 times that of Uranus. Voyager 2 flew past Neptune in 1989 and discovered its atmosphere to be similar in composition and temperature to that of Uranus. Voyager 2 also discovered faint belts and zones, cloud patterns and a great dark spot in the atmosphere of Neptune. However, unlike the Great Red Spot of Jupiter, the dark spot was not evident when the Hubble Telescope was first used to observe Neptune. The presence of the belts and zones on Neptune is thought to be because it is heated internally.

Voyager 2 revealed the presence of a ring system around Neptune and the existence of six more moons in addition to Triton discovered in 1846 and Nereid discovered in 1949. Triton is the largest of Neptune's moons and has a diameter about 0.75 times that of the Earth's moon. Triton orbits Neptune in the opposite direction to the direction in which Neptune spins and its orbit is inclined at 23 degrees to Neptune's equator. Voyager 2 observed geysers spouting high about Triton's icy surface which is wrinkled in parts and smooth in others. Triton is thought to have formed elsewhere in the solar system and then been captured by Neptune in a close encounter. Its geysers could be liquid nitrogen and rock bursting through its surface of nitrogen ice.

see also...

Planet; Planetary Orbits; Uranus

Neutron Star

A neutron star is an extremely dense star composed only of neutrons. Every atom of matter contains a positively charged nucleus consisting of protons which are positively charged and neutrons which are uncharged. Electrons which are negatively charged move about the nucleus of an atom at relatively large distances. In a neutron star, no electrons or protons are present and the entire star consists of neutrons packed together as they are in the nucleus of an atom.

Owing to the fact that neutrons are uncharged particles, they do not repel each other electrostatically like protons do. In 1934, Walter Baade and Fritz Zwicky published a paper in which they advanced the idea of a star composed only of neutrons. They put forward the theory that a supernova is an event that happens when an ordinary star changes into a neutron star. The density of such a star is far greater than the density of a white dwarf. A neutron star with a mass greater than that of the sun would have a diameter of little more than 10 km. The strength of gravity at the surface of a neutron star would be so great that light would be bent by it, almost strong enough to prevent light escaping.

Theoreticians at the time knew that Einstein's General Theory of Relativity predicted the existence of black holes, objects so massive that not even light could escape. Clearly, the prediction of neutron stars turned the black hole from a mathematical prediction into a physical possibility. What experimental evidence exists for neutron stars? In 1967, Jocelyn Bell, a Cambridge research student, discovered a source of regular radio pulses in the sky. Within a year, 20 more similar stars, referred to as 'pulsars', were discovered. Astronomers worked out that a pulsar is a fast-spinning neutron star that emits a radio beam which sweeps around like a lighthouse beam as the star spins. Each time the beam sweeps past the Earth, a pulse of radio waves is detected from the star. The neutron star at the centre of the Crab nebula is a pulsar rotating 30 times each second.

see also...

Black Holes; Pulsar

Newton, Isaac

ir Isaac Newton was born in 1642, the same year that Galileo died, in the market town of Grantham in Lincolnshire. His father died before he was born and he was brought up by a grandparent after his mother remarried. He was sent to the local grammar school as a boarder and entered Cambridge University in 1661. Newton returned home during 1665 and 1666 because the university was closed due to the Great Plague; in those two years he produced mathematical theorems and physical theories including his law of gravitation that revolutionized mathematics and physics. He returned to Cambridge in 1667 and was appointed two years later at the age of 26 to the Chair of Mathematics at Trinity College.

Newton's interests in science were wide-ranging and included astronomy, chemistry and optics as well as mathematics and physics. He set out his theories of mathematics and physics in his greatest work, the *Principia*, in which he showed that his three laws of motion and his law of gravitation are sufficient to explain the motion of any system of bodies. He proved once and for all that the planets and the Earth orbit the sun and he explained Kepler's laws and Galileo's laws. Using his law of gravitation, he was able to predict comets, eclipses and tides. His ideas provided the guiding principles for science for the next two centuries until Einsten showed that space and time are not independent quantities.

Newton published the *Principia* in 1687 and established himself as the foremost scientist of his generation. Sadly, he became involved in a bitter dispute with Leibnitz who claimed to have invented calculus before Newton. At the university, promotion was blocked because he belonged to the Unitarian Church and did not believe in the Holy Trinity. He left the university in 1696 to become Master of the Mint where he devoted his talents to monetary reform. His pre-eminence as a scientist was recognized in 1703 when he was elected President of the Royal Society as well as being knighted.

see also...

Galileo; Kepler's Laws of Planetary Motion; Newton's Law of Gravitation

Newton's Law of Gravitation

Before Newton produced his universal law of gravitation, it was generally believed that objects possessed gravity which pulled down and levity which pushed up. Newton dismissed the concept of levity and assumed that a force of gravitational attraction exists between any two objects. He explained the motion of an object falling to the ground by saying that the object and the Earth attract each other. He used the same idea to explain why the moon goes around the Earth and why the planets go around the sun. If the force of gravity between the sun and the planets suddenly ceased to exist, each planet would continue in uniform motion in a straight line at a tangent to its orbit. The force of gravitational attraction between the planet and the sun keeps the planet circling the sun.

Newton thought that the force of gravity between two objects, imagined as points, was proportional to (1) the mass of each object, and (2) the inverse of the square of the distance between the two objects. For two such point objects of masses m_1 and m_2 at distance apart r, he formed the following equation for the force of gravity F between the two masses.

$$F = \frac{G\, m_1\, m_2}{r^2}$$

where G is a constant which he referred to as the Universal Constant of Gravitation.

Newton's choice of r^2 in his equation rather than r or r^3 or some other power of r was inspired by his previous discoveries of the laws of motion. He had worked out that a body in steady circular motion always experienced an acceleration towards the centre of the circle equal to the (speed)^2/radius. By linking this to his force formula, he proved Kepler's 3rd Law of planetary motion. Any other power of r in his force formula would not have proved Kepler's 3rd Law. Newton's next step was to try to extend his ideas beyond point objects. This turned out to be very difficult and eventually, after many years, he proved that his law of gravitation could be applied to any two objects provided the distance in the equation was the distance between their centres of gravity.

see also...

Newton, Isaac; Kepler's Laws of Planetary Motion

Nova

A nova is a star that suddenly becomes much brighter then fades out gradually, appearing as a new star in the sky if it was previously too faint to be seen. Nova Aquilae appeared dramatically in 1918, within days becoming as bright as Sirius, the brightest star in the sky, and remaining visible to the unaided eye for several months. Like comet-spotting, hunting for novae is one way in which amateur astronomers can make a name for themselves as a nova is an unexpected event and professional astronomers cannot usually afford to observe the sky at random. Peter Collins, an amateur astronomer in Colorado in 1992, was the first person to spot a nova in the constellation of Cygnus. Within hours of its discovery, astronomers world-wide and instruments on the International Ultraviolet Explorer were observing this nova, V1974 Cygni.

A nova is a dramatic event in which a star throws off a shell of matter, brightening by perhaps ten magnitudes or more as a result. The expanding shell of matter is usually too faint to observe directly but its presence is evident from the broad emission lines in the star's spectrum.

So what causes a star to behave in this way? One possible cause is where a white dwarf star draws off matter from another star which is its binary companion. A white dwarf is a very hot collapsed star near the end of its life. Its gravitational pull on the less dense matter of the ordinary star is sufficient to draw such matter towards itself. The extra matter refuels the white dwarf, causing its outer layers to overheat dramatically and, with a great outburst of light, throw off much of its accumulated matter. Some nova have been observed to give a repeat performance. In 1946, T Coronae repeated its performance of 1866 when it flared up by 7 magnitudes to reach magnitude 2. A type Ia supernova is a much more dramatic event in which a white dwarf attracts so much mass from a binary companion that the entire white dwarf overheats and blows itself apart.

see also...

Magnitude; Spectra; Supernova; White Dwarf

60

Nuclear Fusion

The energy radiated from a star is due to the process of nuclear fusion which takes place in the core of the star. In this process, light nuclei are fused together to form heavier nuclei and energy is released in the form of gamma radiation and high energy particles. The energy released per kilogram of hydrogen in this process is about 10 million times greater than the energy released per kilogram when oil burns.

The temperature in the core of a star is of the order of tens of millions of degrees and nuclei move about very fast. The nuclei are positively charged and two such nuclei would repel each other on collision if they were moving slowly. However, because they are moving so fast, two nuclei in collision overcome the force of repulsion and fuse together to form a heavier nucleus. In main sequence stars, hydrogen nuclei,which are single protons, fuse together in several stages to form helium nuclei:

First, two protons fuse together, causing one of the protons to change to a neutron to form a deuterium nucleus consisting of a proton and a neutron. Then the deuterium nucleus fuses with a further proton to form a helium 3 nucleus which consists of two protons and a neutron. Finally, two helium 3 nuclei then fuse together and release two protons, leaving two protons and two neutrons bound together as a helium 4 nucleus.

In a main sequence star, as long as sufficient hydrogen remains in the core, fusion continues and the outflow of radiation prevents the star from collapsing under its own gravity. When the hydrogen in the core is exhausted, the core collapses and the outer layers of the star swell out and cool so the star becomes a red giant. The helium nuclei in the core then fuse stage-by-stage to form heavier nuclei and to release more energy. This process continues until heavier nuclei such as iron nuclei are formed. As energy is needed for the fusion of nuclei heavier than iron, no further fusion occurs and the entire star collapses to form a white dwarf.

see also...

Hertzsprung Russell Diagram;
White Dwarf

Planet

The nine known planets in order of increasing distance from the sun are: Mercury, Venus, Earth, Mars, Jupiter, Saturn, Uranus, Neptune and Pluto. The outermost three planets cannot be seen without the aid of a telescope. The four innermost planets are called the 'terrestial planets' because they are solid and much smaller than Jupiter, Saturn, Uranus and Neptune which are known as the 'giant gas' planets.

The planets are visible because they reflect sunlight. The brightness of a planet changes according to its position on its orbit relative to the Earth because the proportion of its sunlit surface visible from Earth changes; its distance from Earth changes; and, in the case of Saturn, its rings reflect more sunlight when they are seen 'face on' than when seen 'edge on'.

The planets gradually move through the constellations near the ecliptic as they orbit the sun. The word planet meant 'wanderer' in Ancient Greece as a planet changes its position against the fixed background of the stars in the constellations.

Mercury and Venus rise and set within a few hours of sunrise or sunset because they are closer to the sun than the Earth, and they show a complete cycle of phases like the moon. At 'inferior conjunction' when moving between the Earth and the sun, Venus can be seen as a large crescent as its sunlit surface is facing mostly away from the Earth. As the planet moves around the sun the fraction of its sunlit surface increases until we see its full sunlit surface as a relatively small disc when it is near 'superior conjunction' almost diametrically opposite the Earth on its orbit.

Mars and the other planets further from the sun do not show a complete cycle of phases as they never pass between the Earth and the sun. An outer planet is said to be at 'opposition' when it is in the opposite direction to the sun from the Earth and therefore can be seen by observers in the northern hemisphere due south at midnight.

see also...

Earth; Jupiter; Mars; Mercury; Neptune; Pluto; Saturn; Uranus; Venus

Planetary Orbits

The orbit of a planet is its path around the sun. The planets orbit the sun in the same direction and almost in the same plane as each other. The force of gravity due to the sun makes a planet or a comet keep going around the sun on the same path. In general, the orbit of a planet or comet around the sun is egg-shaped or elliptical, with the sun at one of the focal points of the ellipse. The was first deduced as a result of observations by Johannes Kepler in the sixteenth century. Fortunately, the Earth's orbit is almost circular. If it was not, the Earth would experience a much greater range of temperatures each year. Pluto is on an orbit which brings it closer to the sun than its nearest neighbour, Neptune, for a small fraction of its orbit. The diagram below shows how to draw an ellipse.

The orbit of a planet is characterized mainly by its mean radius and its time period. The mean radius is half the average of the maximum diameter and the minimum diameter. The time period of a planet is the time it takes to move exactly once around the sun. The further a planet is from the sun, the longer the time period of the planet is. This is because the distance around the orbit is longer and also because it moves more slowly. Kepler deduced from his observations that the square of the time period of a planet is in proportion to the cube of its mean radius. This is known as Kepler's 3rd Law. For example, Saturn has a time period of 29.4 years and a mean orbital radius 9.5 times that of the Earth. Check for yourself that 29.4 squared is equal to 9.5 cubed to within 1 per cent. Kepler's 3rd Law can be explained using Newton's law of gravity and Newton's laws of motion.

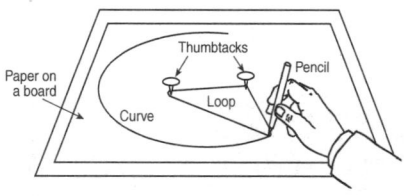

Drawing an ellipse

see also...

Kepler's Laws of Planetary Motion; Planet

Precession of the Equinoxes

The Earth spins at a steady rate about an axis that points in a direction which scarcely changes. This is why the Pole Star is directly above the Earth's North Pole. The Earth's spin axis is tilted at an angle of 23.5 degrees to its orbital axis which means that the Equator is 23.5 degrees to the ecliptic. Imagine a model of the Earth and the Sun in which the Earth's Equator was in the same plane as its orbit. The Earth would be a very boring place as every day would consist of 12 hours of daylight and 12 hours of darkness. In our model, the Earth would need to be tilted by 23.5 degrees to represent the real Earth on its orbit around the sun.

next millennium, in the year 3000, the Earth's axis will have turned through about 35 degrees. This effect, known as 'precession', is observed when a child's spinning top is tilted, making its axis slowly turn about a vertical line through its pointed base. The Earth is precessing at a rate of once every 26,000 years. The effect of precession is to make the equinoxes, the points where the Celestial Equator crosses the ecliptic, move gradually along the ecliptic at a rate of about one degree every 70 years. The precession of the equinoxes is the reason why the First Point of Aries, which is the position of the vernal equinox on the Celestial Sphere, is now in Pisces, not Aries where it was 3000 years ago when the vernal equinox was named as the First Point of Aries.

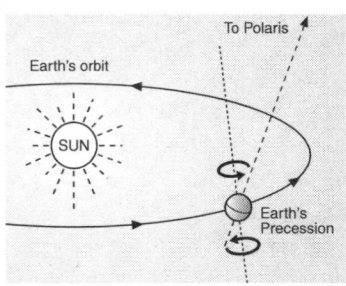

Precession of the Earth's spin

The direction in which the Earth's axis points is gradually changing. By the

see also...

Celestial Sphere 2

Proper Motion

The stars in a constellation form a recognizable pattern which is the same now as it was centuries ago. Stars within 100 parsecs of the sun change position over six months then change back again over the next six months. This effect is caused by **parallax** due to the Earth's motion around the sun. However, some stars shift position against the fixed star background over a period of years. Photographs of such a star and its neighbours would show the mean annual position of the star gradually changing. This effect is known as 'proper motion' and is due to the star's motion relative to the Sun and its comparative nearness. Barnard's Star has the largest known proper motion, changing its position gradually at a rate of about 0.3 degrees per century. This star is a magnitude 9 red dwarf just 6 light years from the sun, moving through space at over 160 km/s. If it were 600 light years away, its proper motion would be much smaller at 0.003 degrees per century. Clearly, a star that shows no proper motion is too far away to change its position in its constellation by a measurable amount although it may be moving faster than Barnard's Star. The stars that make up the fixed background of stars in a constellation are therefore too far away to show proper motion although present records compared with measurements taken many centuries ahead might show the effect for such a star.

The proper motion of a star is used to calculate the speed and direction of the star relative to the sun. This can be determined if its radial speed which is the distance it moves per second towards or away from the sun and its tangential speed, the distance it moves per second perpendicular to its line of sight, are measured. The radial speed is measured from the doppler shift of its line spectrum of light. Its tangential speed is calculated by multiplying its distance from the sun by its proper motion in radians per second (where 1 radian = 180 /π degrees). Knowledge of the speed and direction of stars near the sun relative to the sun has enabled the sun's own speed and direction to be calculated, giving a result of 4.2 AU per year in the direction of Hercules.

see also...

Constellation 1; Parallax; Spectra

Ptolemy's Planetary Model

Ptolemy was an Egyptian astronomer in the second century AD who wrote a great work or *Amalgest* consisting of 11 books covering everything known at that time about astronomy. His catalogue of 1022 stars was not surpassed for three centuries. He is best remembered for his model of the universe which placed the Earth at the centre with each planet rotating on an 'epicycle', a circle whose centre circles the Earth. (See below.)

Ptolemy believed that the sun and moon moved around on circles in the same plane centred on the Earth; the moon's orbit being much smaller than the sun's. He also thought that the stars were fixed to the Celestial Sphere which encompassed all the planets. The Celestial Sphere spun at a steady rate of once every sidereal day, taking the stars that rose each night across the night sky from east to west. Mercury and Venus moved around on epicycles centred on the line from the Earth to the sun, with Mercury closer to the Earth than Venus. As the sun moved around the Earth, Ptolemy showed why Mercury and Venus were never far from the sun. Each of the other planets moved independently of the sun on its own epicycle which moved around the Earth on its own circle. The radial line from an outer planet to the centre of its epicycle was always parallel to the line from the Earth to the sun. Using his model Ptolemy was able to explain the retrograde motion of Mars, Jupiter and Saturn.

The Celestial Sphere

Ptolemy's model

see also...

Copernicus

Pluto

Pluto was discovered in 1930 by Clyde Tombaugh using a wide-angle camera fitted to a telescope. Tombaugh found that the planet is much fainter than Neptune and moves much more slowly through the constellations. Careful measurements on its progress have enabled astronomers to work out that Pluto moves around the sun on a much more elliptical orbit than any other planet, its distance varying between 30 and 50 AU, taking 248 years for each orbit. From 1979 to 1999, Pluto was closer to the sun than Neptune. The inclination of its orbit to the ecliptic at 17 degrees is also much larger than that of any other planet. Its diameter is less than 0.2 times that of the Earth and its mean density is 2.3 times that of water.

In 1978, Pluto was discovered to be accompanied by a moon, orbiting Pluto once every six days at a distance of about 20,000 km. This object, named Charon by its discoverer James Christy, is locked into synchronous orbit and rotation with Pluto. For a few years, Charon and Pluto eclipsed each other regularly when the orbital plane of Charon happened to be edge-on to the Earth.

Pluto is thought to have a low-density atmosphere of nitrogen and carbon monoxide. More information about Pluto and Charon has been obtained using the Hubble Space Telescope. Bright polar ice caps and bright and dark regions which rotate near Pluto's equator indicate its six-day rotation period and also show that Pluto's spin axis is tilted by about 120 degrees to its orbit. Charon has a diameter about half that of Pluto so the two objects are more similar in size than any other moon and its parent planet. They have similar density and chemical composition and both surfaces are partly covered in ice.

see also...

Neptune; Planet; Planetary Orbit

67

Pulsar

A pulsar is like a cosmic lighthouse, emitting beams of light or radio waves that sweep around the sky. The radio beam from a pulsar repeatedly sweeps past the Earth, causing a pulse of radio-wave energy to hit the Earth each time. Pulsars were first detected in 1967 when radio astronomers discovered regular bleeps from space. At the time, some newspapers claimed that intelligent life beyond the Earth was trying to communicate.

A pulsar is a spinning neutron star that emits beams of electromagnetic radiation that sweep around as it spins. The first pulsar was found to produce radio pulses every 1.33 seconds, corresponding to a time period of rotation of 1.33 seconds. In comparison, the sun takes about four weeks to rotate once. If the sun shrank in size without loss of mass, it would spin faster, just as an ice skater does when she pulls her arms in. The fact that pulsars are spinning very fast made astronomers realize that pulsars must be very small in comparison with the sun and therefore very dense. A neutron star of mass equal to that of the sun

would be no more than about 20 km in diameter and would spin much faster than the sun. Further pulsars were soon discovered, including a 33 millisecond pulsar at the centre of Crab Nebula. This pulsar was the first pulsar to be found emitting optical pulses. This discovery supported the theory that a pulsar is a neutron star which is the remnant of a supernova explosion.

A pulsar has an extremely strong magnetic field (of the order of 10^6T) with poles aligned on an axis at a non-zero angle to its rotation axis. The magnetic field guides electromagnetic radiation into beams along the magnetic axis which means that the beams sweep around as the star spins. The pulse rate of a pulsar decreases slowly, indicating that its rate of rotation decreases gradually as it loses energy. The period of the Crab pulsar is increasing by about ten microseconds per year. In general, the older a pulsar is, the slower its rate of rotation.

see also...

Evolution of Stars; Neutron Star; Radio Astronomy

Quasar

A quasar is an astronomical object as bright as a galaxy but much smaller, like a star. 'Quasar' is short for 'quasi-stellar object' which means an object like a star. Quasars are billions of light years away yet a quasar is bright enough to be seen even though it is about the same size as a star. A star the same distance away would be much too faint to see.

The first quasar was discovered in 1962 when a previously discovered astronomical radio source referred to as 3C 273 was identified as a thirteenth magnitude star with a red shift of 0.15, corresponding to a speed of recession of 15 per cent of the speed of light and a distance of over 2000 million light years away. It was calculated to be over 1000 times more luminous than the Milky Way galaxy, yet records of its light output showed variations from year to year. Such a timescale corresponds to the time taken by light to travel across the object, thus indicating its size to be no more than that of the order of light years. In contrast, the disc of the Milky Way galaxy is over 100,000 light years in diameter. Many more quasars have subsequently been detected with red

shifts between about 1 and 4, corresponding to distances between 5000 and 10,000 million light years away and recession speeds in excess of 85 per cent of the speed of light. This recession is due to the expansion of the universe after the Big Bang. Few quasars lie beyond red shift 4 indicating a quasar age between 5000 and 10,000 million years ago, commencing about 2000 million years after the Big Bang. Quasars are among the oldest and most distant objects in the observable universe and there is no evidence of recent nearby quasars.

Detailed radio images of quasars indicate the presence of fast-moving clouds and jets of matter being ejected. The presence of jets of matter may be due to the presence of a massive black hole at the centre of a galaxy early in its formation. A large galaxy with a massive black hole at its centre would have destroyed any smaller galaxies which collided with it, sweeping such galaxies forever into the black hole at its centre.

see also...

Big Bang; Black Holes; Magnitude; Red Shift

Radar Astronomy

Radar is an acronym for **RA**dio **D**etection **A**nd **R**anging which is the use of pulsed short-wavelength radio waves to detect objects that reflect the waves and to measure the distances to such objects. Radio telescopes are used to direct such radar pulses at the moon or at rocky planets such as Mars and Venus and to detect the reflected pulses. By timing the interval between the transmission of a pulse and the detection of its reflection from a distant object, the distance to the object can be calculated by multiplying half the round-trip time by the speed of electromagnetic waves in space. For example, a round-trip time to Mars of 42 minutes would give a distance equal to 21×60 seconds $\times$ 300,000 km/s = 378 million km, using the value of 300,000 km/s for the speed of electromagnetic waves in space.

Large radio telescopes used to detect radio waves from space are used for radar astronomy. Radar waves are short-wave radio waves in the wavelength range of the order of centimetres. Radar pulses transmitted from a large parabolic dish spread out little before reaching the distant object but they are scattered when reflected from its surface so the detector must be extremely sensitive as the reflected pulses are extremely weak. Nevertheless, the accuracy of such measurements has enabled the diameter and the surface topography of a planet to be determined by scanning the pulses across the object's surface. In addition, the measurement of the doppler shift of the reflected pulses from opposite ends of the planet's equatorial diameter has enabled the planet's speed of rotation to be measured precisely.

Radar astronomy has also been used to confirm that electromagnetic waves grazing the sun are deflected by the sun's gravity, in accordance with predictions from Einstein's *General Theory of Relativity*. The effect is to increase the round-trip time of radar pulses to and from a planet as the line of sight to it from Earth approaches closer and closer to the sun. This increase was measured and found to agree with the predicted increase to within 0.1 per cent.

see also...

Planet; Radio Telescope; Red Shift

Radio Astronomy

R adio astronomy is the detection and measurement of astronomical sources of radio waves which are electromagnetic waves in the wavelength range from a few centimetres upwards. The Earth's atmosphere allows radio waves of wavelengths less than about ten metres to reach the ground. Large radio telescopes can therefore be used to map out radio sources in the sky. The sun was found to be a strong source of radio waves by scientists in 1942 working on wartime radio communications. In 1946, a strong radio source in Cygnus referred to as Cygnus A was discovered. Further radio sources were found, such as the Crab Nebula, now known to be a supernova remnant, and the galaxy M87. In addition, radio emission was detected from the Milky Way, indicating radio sources in the disc of the Milky Way galaxy.

The distribution of hydrogen in the disc of the Milky Way galaxy was mapped out using radio telescopes tuned to detect 21 cm wavelength radio waves. Such radio waves are emitted from a hydrogen atom when its electron, with its spin parallel to the spin of the

proton, flips to a lower energy state with its spin in the opposite direction. Unlike light, radio waves pass through the dust clouds that obscure much of the disc of the Milky Way galaxy. Radio waves from hydrogen gas clouds in the Milky Way disc were mapped out and, by measuring their doppler shifts of wavelength, the motion and distribution of the gas clouds was deduced, providing a map of the spiral arms of the Milky Way galaxy. Clouds of molecules such as ammonia and carbon monoxide have been discovered in the plane of the galaxy as a result of detecting strong signals at certain radio wavelengths known to be due to such molecules.

Many more radio sources have since been detected including pulsars, quasars and supernovae. The discovery that the distribution of extra-galactic radio sources increases with distance created doubts about the Steady State Theory of the universe and led to the discovery of quasars.

see also...

Electromagnetic Radiation; Pulsar; Quasar; Radio Telescope; Red Shift; Supernova

Radio Telescopes

The single dish steerable radio telescope consists of a large parabolic reflecting dish with an aerial at its focal point. When the dish is directed at a source of radio waves in the sky, radio waves reflect from the dish onto the aerial to produce a signal which is a measure of the intensity of the waves emitted by the source. The signal from the aerial is supplied to a high-gain amplifier which strengthens the signal and feeds the strengthened signal to a computer for analysis and recording. The dish usually consists of a wire mesh which is lighter than metal sheets and just as effective as a reflector of radio waves, provided the mesh spacing is less than one-twentieth of the wavelength of the radio waves being detected.

The amplifier must make the signal from the source stronger without amplifying the background noise or 'hiss' which consists of random variations due to local radio sources and to the random motion of electrons in the amplifier components. By averaging the signal out over successive short intervals, background noise is eliminated.

The dish diameter determines the collecting area of the dish so that fainter and fainter sources can be detected using bigger and bigger dishes. The dish diameter also determines the resolution of the telescope which is a measure of the detail it can map out. Two radio sources close together would be mapped out as a single source if the dish diameter was too small, as diffraction from the dish would spread out the image of each source too much. The Lovell Radio Telescope in Cheshire, England has a 78 m dish which gives a resolution of 0.2 degrees for 21 cm wavelength radio waves. The Arecibo Radio Telescope in Puerto Rica is a 300 m fixed concave dish set in a natural bowl.

Much improved resolution is achieved by linking radio telescopes together. In effect, the resolution is then equivalent to a single dish of diameter equal to the distance between the radio telescopes, although only along a line parallel to the line between the telescopes.

> ### see also...
> Radio Astronomy; Telescopes 3

Red Giant

A giant star is a star that is much larger than the sun. If the surface temperature of such a star is cooler than the sun, its colour is orange or red rather than yellow and thus the star is referred to as a 'red giant'. A red giant has an absolute magnitude of about zero or a negative absolute magnitude so its position on the Hertzsprung Russell diagram lies high above the Main Sequence. Such a star emits at least 100 times as much light as the sun because the difference in absolute magnitudes is about 5; however, the sun's temperature is twice as high as that of a red giant so the sun emits 16 times as much light per unit area (as the light emitted is in proportion to the fourth power of the temperature). Thus a red giant has a surface area at least 1600 times larger than that of the sun and is therefore about 40 times larger in diameter than the sun. Arcturus in Bootes is an orange-red star with an apparent magnitude of about –0.1 at a distance of about 37 light years from the sun. It can be seen by observers in the northern hemisphere due south before midnight in May. Its absolute magnitude is –0.4 and its diameter is about 40 times that of the sun.

Stars that are about 5 or more magnitudes higher than red giants on the HR diagram are known as 'super giants' as they are up to 300 times larger in diameter than the sun. Antares, the largest supergiant, is an M class star with an apparent magnitude of 0.9 at a distance of 520 light years from the sun. Its absolute magnitude is –5.1 and its diameter is about 300 times that of the sun.

A red giant or a supergiant is a star that has left the Main Sequence after the collapse of its core and the swelling and cooling of its outer layers. This occurs when all the hydrogen in the core has been converted into helium. When all the helium in the core has been converted into heavier elements, a giant star collapses and becomes a white dwarf or a supernova.

see also...

Evolution of Stars; Hertzsprung Russell Diagram; Magnitude; Thermal Radiation

Red Shift

The light spectrum from a star or a galaxy is a continuous spectrum crossed by dark vertical lines at wavelengths characteristic of elements in the outer layers of the star. The lines of the spectrum of light are shifted due to the star's motion if it is moving towards or away from us. This is an example of the doppler effect which is the change of the observed wavelength of waves emitted from a source in motion relative to the observer. The spectral lines are shifted to longer wavelengths (i.e. a red shift) if the light source is receding and to shorter wavelengths if the light source is approaching (ie. a blue shift). For light emitted from a monochromatic light source of frequency f which is moving at speed υ, it can be shown that the shift of wavelength $\Delta\lambda = \upsilon/f = (\upsilon/c)\,\lambda$, where c is the speed of light and λ is the wavelength of light. Thus the speed of a distant star or galaxy can be measured from the wavelength shift $\Delta\lambda$ of its spectrum using the equation $\upsilon = c\,\Delta\lambda/\lambda$.

By observing the spectrum of light from different galaxies using the 60 cm telescope at the Lowell Observatory, Vesto Slipher in 1917 discovered that certain spiral galaxies are moving away from us at speeds of more than 500 km/s, much faster than any object in our own galaxy. The term 'red shift' was coined for the ratio of the change of wavelength to the emitted wavelength. Thus a red shift of 0.1 means that the source is receding at a speed of 0.1 c. Edwin Hubble followed up Slipher's work by estimating the distances to two dozen galaxies of known red shifts, thus discovering Hubble's Law which is that the speed of recession of a distant galaxy is in proportion to its distance.

In 1963, Maarten Schmidt discovered the first quasar as a result of finding that the spectral lines of a star-like object 3C 273 are red-shifted by about 15 per cent. He deduced that this object is receding at about 0.15 c and must be over 2 billion light years away and therefore much more powerful than an ordinary star. Many more quasars have since been discovered.

see also...
Hubble's Law; Quasar; Spectra

Retrograde Motion

The retrograde motion of an outer planet is the reversal of its progress through the constellations of the night sky for a period of time from before opposition to after opposition. An outer planet normally moves gradually from west to east through the constellations, appearing on successive nights to be a little further eastwards relative to the constellation it is in. The eastwards movement is because the planet moves gradually around the sun on its orbit counter-clockwise as seen from the northern hemisphere. However, because the planet moves more slowly around its orbit than the Earth does, the Earth on its orbit nearer the sun catches up with the planet and overtakes it at opposition. As the planet approaches opposition, the effect of the Earth catching up with it is to halt then reverse its eastward progress. When the Earth has moved sufficiently far beyond opposition, the reversal is halted and normal eastwards progress is resumed.

The retrograde motion of Mars occurs once every 26 months, lasting two to three months in total. Because the orbit of Mars is not circular, the most favourable opposition of Mars occurs when its distance to Earth at opposition is least. Jupiter is at opposition once every 13 months and its retrograde motion is less evident than for Mars because its distance from Earth at opposition is about eight times greater. Ptolemy, an Egyptian astronomer who lived in the second century AD, devised a model of planetary motion which successfully explained the retrograde motion of the outer planets in terms of epicycles. Ptolemy's model provided the accepted explanation of the motion of the planets for over 14 centuries until it was rejected in favour of the Copernican model of the solar system.

see also...

Copernicus; Planetary Orbits; Ptolemy's Planetary Model

Saturn 1 – The Planet and its Moons

Saturn has a diameter which is 9.5 times that of the Earth and is the second largest planet in the solar system. Its mean distance from the sun is 9.5 AU and it orbits the sun once every 29.5 years. Its ring system is in its equatorial plane, extending out to a diameter over twice as large as the planet's diameter. The rings reflect sunlight, causing Saturn to be at its brightest when its rings are face-on. In addition, telescopic observations of Saturn reveal the presence of bright and dark belts parallel to its equator, although less distinct than those of Jupiter.

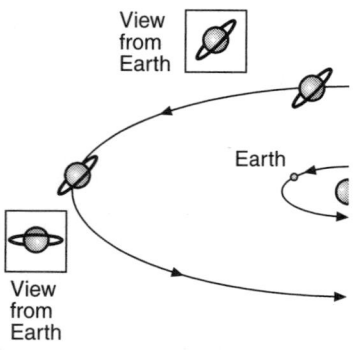

Saturn seen from the Earth

Saturn rotates once every 10 hours, causing its equatorial diameter to be about 20 per cent greater than its polar diameter. The planet is thought to have an atmosphere composed of 90 per cent hydrogen and 10 per cent helium, at a temperature of −180° C, which lies above a sphere of liquid hydrogen with a relatively small solid core at its centre.

Titan, the largest of Saturn's 18 known moons, orbits the planet once every 15 days at a mean distance ten times greater than Saturn's diameter. Its atmosphere, thought to be predominantly nitrogen and methane, is extremely cold at about −180° C and is covered by a dense orange smog layer. Except for Tethys, Dione, Rhea and Iapetus, all the other moons have diameters much less than a quarter of the diameter of Titan. Iapetus is known to have one hemisphere of its surface much brighter than the other hemisphere. Its darker side faces the direction in which it is moving, so Iapetus can only be seen from Earth when it is west of the planet moving away from Earth.

see also...

Planet; Planetary Orbits

Saturn 2 – The Rings of Saturn

Without the aid of a telescope, Saturn appears as a yellow star moving gradually through the constellations near the ecliptic. Using a telescope, its rings may be seen facing the Earth once every seven years. Its rings remain in the same plane as it moves around the sun, tilted by 26 degrees to its orbit. Seen from the Earth, its rings appear in different orientations, as shown below, according to the position of Saturn on its orbit. When the rings are edge-on, they can be seen only through a powerful telescope because they are very thin.

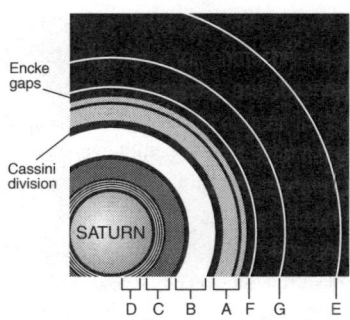

Saturn's rings in close up

Saturn's rings as seen from Earth consist of a faint outer ring, known as the A-ring, separated by a prominent gap known as the Cassini division from a bright inner ring known as the B-ring. A further much fainter ring, the C-ring, is present inside the B-ring. The A-ring is divided by a narrow gap known as the Encke division. The Cassini division and the Encke division exist due to gravitational effect of the moons of Saturn in orbit just beyond the rings which causes any particles in the two gaps to be swept away.

Further rings include rings labelled D, E, F and G. The D-ring is extremely faint and it lies between Saturn's atmosphere and the C-ring. The F-ring lies beyond the A-ring between two moons, Pandora and Prometheus. The G-ring is very faint and lies further out than the F-ring just within the orbit of Mimas. The E-ring is also very faint and is along the orbit of Enceladus. Voyager 2 provided detailed pictures of the rings, confirming that they are extremely thin in comparison with their diameters, and they are composed of particles ranging in size from dust particles to large boulders.

see also...

Saturn 1

Sidereal and Solar Time

Our daily lives are governed by solar time. 'One solar day' is defined as the time between successive transits of the sun across the inferior meridian which is the half of the meridian below the horizon. The solar day is from midnight to the next midnight which is the same duration of time as from midday to the next midday. The solar day is divided into 24 hours.

The Earth spins at a steady rate, causing each star except Polaris to move across the night sky. The progress of a star across the sky is like the progress of the hour hand of a clock, travelling round once per day. The time taken by a star from a transit of the meridian to its next transit of the meridian is defined as 'one sidereal day'. Sidereal means according to the stars. The sidereal day is 23 hours and 56 minutes in terms of the solar day. This is because the Earth progresses around its orbit of the sun so the constellations gradually change in the night sky. A non-circumpolar star rises about four minutes earlier each night due to the fact that the Earth moves about one degree per day along its orbit, causing a star to rise about four minutes earlier each night. Over a period of a month, therefore, a star rises about two hours earlier.

Astronomical observatories usually have a clock that keeps sidereal time next to a clock that keeps solar time. The sidereal day commences when the First Point of Aries crosses the (superior) meridian at the Observatory. The right ascension of a star is the interval of time (in sidereal hours) from the transit of the First Point of Aries across the meridian to the transit of the star. Hence a star crosses the meridian of an observatory when the time on the sidereal clock at the observatory is equal to the star's right ascension. Atomic clocks are used for accurate timekeeping because the Earth's rate of rotation varies slightly. The second, the scientific unit of time, is defined in terms of the frequency of vibrations of a certain type of atom in an atomic clock. Leap seconds are periodically added or removed to keep atomic time in step with solar time.

see also...

Celestial Sphere 3

Solar System

The Solar System consists of the sun, the planets and their moons, asteroids and comets. Some of the physical properties of the planets are given in the table below. The sun is included for comparison.

orbiting the sun in the same direction and in the same plane as each other. The heavier materials such as iron and silica in the rotating disk settled close to the sun to form the inner planets while the lighter elements including ice remained in the outer

	Mean distance from the Sun in AU	Period in years	Diameter in Earth diameters	Density relative to water
Sun	–	–	109.00	1.40
Mercury	0.39	0.24	0.38	5.40
Venus	0.72	0.61	0.95	5.20
Earth	1.00	1.00	1.00	5.50
Mars	1.52	1.88	0.53	3.90
Jupiter	5.20	11.90	11.20	1.30
Saturn	9.50	29.40	9.50	0.70
Uranus	19.20	83.80	4.00	1.30
Neptune	30.10	163.70	3.90	1.80
Pluto	39.50	248.00	0.18	1.10

The sun formed about 4500 million years ago from a slowly rotating cloud of interstellar gas and dust which contracted under its own gravity. The central region of the contracting cloud became denser and hotter, forming a so-called protostar which continued to contract and heat up until nuclear fusion commenced at the core. The outer regions of the rotating cloud formed a rotating disk from which the planets formed,

region of the disk to form the outer planets. Huge solar flares from the newly formed sun as it stabilized itself would have driven extra-planetary gas and dust far beyond the planets, perhaps leading to the formation of the Oort cloud which is where long-period comets are thought to originate from.

see also...

Asteroids; Comets; Planet; Stars 2

Solar Wind

The solar wind is evident when a comet with a visible tail is observed. The tail always points away from the sun, regardless of the direction in which the comet moves. The tail is forced to point away from the sun because of the effect of charged particles such as protons and electrons, referred to collectively as plasma, that stream from the solar corona at speeds of the order of several hundreds of kilometres per second. Space probes have measured the strength of the solar wind at different distances from the sun in terms of the types of particles, their concentration and their speeds. Images from SOHO, a spacecraft designed and used to study the sun, have revealed holes in the solar corona which provide escape channels for the particles of the solar wind. Particles escaping through the permanent holes above the sun's pole travel almost twice as fast as particles escaping from equatorial holes.

The solar wind plasma conducts heat extremely well which causes it to become hotter further from the sun and to accelerate as it moves away from the sun. The motion of the charged particles of the solar wind generates a magnetic field which in effect extends the magnetic field of the sun far into space. The rotation of the sun causes the lines of force of the extended magnetic field to spiral outwards from the sun. Solar flares eject charged particles into space at very high speeds, adding to the steady stream of charged particles from the corona.

The Earth's magnetic field traps charged particles of the solar wind in two doughnut-shaped belts, known as the Van Allen belts, around the Earth. The inner belt extends from a height of about 2000 km to about 5000 km. The outer belt extends from about 12,000 km to about 20,000 km above the surface. At the Earth's orbit, the concentration of protons in the solar wind varies by a factor of several hundred and their speeds range from 300 km/s to 700 km/s. These variations storm through space, at times causing disturbances in the Earth's magnetic field which severely affect radio communications systems.

see also...

Comets; Sun 2

Spectra

The light from a star contains a continuous spectrum of colours. The spectrum of sunlight is evident in a rainbow or by passing a beam of sunlight through a prism and observing the exit beam from the prism on a screen. In both cases, a continuous spread of colours from red to orange to yellow to green to blue and violet are seen. Using a spectroscope, a scientific instrument designed to spread a beam of light into its constituent colours, the spectrum of any light source may be observed. Each colour in a spectrum corresponds to light of a certain wavelength which ranges from about 0.0004 mm for blue light to about 0.0007 mm for red light.

A filament lamp also gives a continuous spectrum but a vapour lamp, such as a sodium lamp, or a discharge tube, such as a neon tube, give a spectrum consisting of bright lines of different colours. The pattern of colours and hence the wavelengths present is characteristic of the type of atoms present in the light source and is referred to as a line emission spectrum. By measuring the wavelength of each colour in a line spectrum, the chemical elements present in the light source can be identified as each type of atom is an atom from a particular chemical element.

The solar spectrum contains dark vertical lines which are seen against the background of the continuous spectrum. These absorption lines occur at certain wavelengths and they are due to certain colours of light from the solar photosphere being absorbed by gases in the outer regions of the sun, such as the solar corona. The pattern of absorption lines is like the pattern of the lines of a line emission spectrum as it may be used to identify the chemical elements present in the absorbing regions. Helium was discovered by Norman Lockyer in 1868 as a result of observing and measuring the line spectrum of sunlight.

see also...

Sun 2

Stars 1 – Surveying the Stars

The sun is a typical star. Stars vary in size from dwarf stars much smaller than the sun to giant stars much larger than the sun. Stars appear in the night sky as points of light because they are so far away. The light from a star provides information on its distance, its velocity, its chemical composition, its surface temperature, its radius, how much radiation energy per second it releases, its mass and its lifetime.

Its distance can be measured directly if it is sufficiently close to us. This is done by the parallax method which involves measuring the angle which it shifts through over a period of six months. The distance to a star beyond about 300 light years cannot be measured by parallax. Its absolute magnitude can be determined and its distance found if its position on the Hertzsprung Russell diagram is known.

Its velocity is determined by measuring the radial and tangential components of its velocity. The doppler shift of the lines of its spectrum need to be measured to enable the radial component of its velocity to be calculated. The tangential component of its velocity can be determined if its distance and proper motion are known.

The chemical composition of a star is determined by measuring the wavelengths of the lines of its line spectrum. These wavelengths are characteristic of the types of atoms that emit the light and hence can be used to identify the chemical elements present.

The surface temperature of a star is determined from its spectral type (i.e. colour). For example, a red star is an M-class star and therefore its surface temperature is about 3000 K. A more precise determination requires the intensity of its spectrum to be measured at different wavelengths to find the wavelength corresponding to its peak intensity. Its temperature is then calculated using Wien's law.

see also...

Distance Measurement 1 & 2;
Hertzsprung Russell Diagram;
Magnitude; Proper Motion;
Thermal Radiation

Stars 2 – Classification

tars differ in colour as well as in brightness. Betelgeuse in the constellation Orion is a red giant star. Rigel in the same constellation is a blue giant star. The spectrum of light from a star is a continuous spectrum of colour from red to orange to yellow to green to blue to violet. The continuous spectrum is crossed by dark absorption lines due to different types of atoms in the outermost layers of the star absorbing light emitted at certain wavelengths from inner layers of the star.

The intensity of each part of the spectrum changes with colour and is

the spectrum. The colour of a star, therefore, is determined by its surface temperature. Stars are classified therefore according to colour or temperature and the absorption lines present. This classification system was first established before the exact link with temperature was known by assigning a different letter of the alphabet to each colour. When the exact link became known as a result of experiments using laboratory sources of light at different temperatures, the order of the letters had to be rearranged from exact alphabetical order to the order shown in the table, so as to match the temperature sequence.

Colour	blue	blue white	white	blue green	yellow	orange	red
Temperature (in thousands of degrees)	30	20	10	8	6	4	3
Star classification	O	B	A	F	G	K	M
Strong absorption lines due to the presence of:	helium	hydrogen, helium	hydrogen light metals	light metals	light metals	metals	metal oxides

most intense at a colour that depends on the temperature of the surface of the star. The hotter a star is, the nearer the peak intensity of its spectrum is towards the blue end of

see also...

Hertzsprung Russell Diagram; Luminosity; Magnitude; Red Giant; Spectra; Thermal Radiation

Stars 3 – Dwarfs and Giants

The luminosity or light energy per second emitted by a star depends on its surface temperature and its surface area in accordance with Stefan's law of radiation (stated opposite). If its surface temperature and radius are known, its luminosity can be calculated. The radius of the sun can be calculated from its distance to Earth and its angular width in the sky. Its surface temperature is known from its colour to be 6000 K. Hence, the energy per second emitted by the sun has been calculated at 400 million million million million watts.

The luminosity of any other star can be calculated if its absolute magnitude is compared with that of the sun. The difference between its absolute magnitude and that of the sun tells us how much energy it emits per second in terms of the energy emitted per second by the sun. For example, if a star is 5 magnitudes more powerful than the sun, it must emit 100 times as much energy per second. Hence, it emits energy at a rate of 40,000 million million million million watts.

If the luminosity and the surface temperature of a star is known, its surface area and hence its radius can be calculated using Stefan's law which states that the energy per second emitted per square metre of the star's surface is proportional to the fourth power of its surface temperature. Thus each square metre of the surface of a star that is half as hot as the sun must emit one sixteenth of the energy per second as the sun emits from each square metre of its surface. If such a star emits 100 times as much energy per second as the sun, its surface area must be 1600 times larger, making its radius 40 times greater than the sun's. Such a star is referred to as a red giant because its colour is red and its diameter is much greater than the diameter of the sun. In the same way, a star that is twice as hot as the sun but is less powerful, can be shown to be much smaller in diameter than the sun. Such a star is referred to as a dwarf star. For example, a star that is 5 magnitudes less powerful than the sun and is twice as hot would be 40 times smaller in diameter than the sun.

see also...

Luminosity; Magnitude; Red Giant; Thermal Radiation

Stars 4 – Mass and Lifetimes

The mass of a main sequence star can be determined from its luminosity in accordance with the relationship between mass and luminosity discovered by Sir Arthur Eddington. By studying binary stars, Eddington was able to show that the luminosity of a main sequence star is approximately in proportion to the cube of its mass.

The mass of the sun is known to be 2×10^{30} kg, a result obtained by applying Newton's theory of gravitation to the motion of the Earth about the sun. To find the mass of the two stars in a binary system, the separation of the two stars and the orbital period needs to be known, and Kepler's 3rd Law in the form below is used:

Mass (in solar masses) $\times$ Period (in years)^2 = Separation (in AU)^3

The individual star masses can be worked out from the total mass, as the mass ratio of the two stars is in inverse ratio to their radii of orbit.

The lifetime of a star depends on its mass because stars are composed mostly of hydrogen which is the fuel of a star. Protons (i.e. hydrogen nuclei) are fused together in the core of a star to form helium nuclei, releasing energy in the process. This fusion process causes energy to be released at a rate of 70 million million watts for every kilogram of hydrogen used each second. Because the sun emits energy at a rate of 4×10^{26} watts, it therefore follows that it converts hydrogen to helium at a rate of 6×10^{11} ($= 4 \times 10^{26} / 70$ million million) kilograms per second. The total mass of the sun is 2×10^{30} kilograms therefore its hydrogen fuel will last for about 3.5×10^{18} seconds which is about 10,000 million years. Thus, the sun will last about 10,000 million years. For a star of mass M in solar masses and luminosity L in terms of solar luminosity, its lifetime is therefore M/L times the lifetime of the sun. Because the luminosity of a main sequence star is approximately in proportion to the cube of its mass, it follows that the greater the mass of a star, the shorter its lifetime.

see also...

Binary Stars; Kepler's Laws of Planetary Motion; Luminosity; Newton's Law of Gravitation

Strength of Gravity

The strength of gravity or gravitational field strength due to a massive object such as a star or a planet is defined as the force of gravity per unit mass on a small object in the gravitational field of the star or planet. The strength of gravity varies in magnitude and direction according to position. For example, the gravitational field strength of the Earth at a height of 1000 km from the Earth's surface is 7.5 newtons per kg in comparison with 9.8 newtons per kg at the surface.

The gravitational field strength near a star or a planet depends on the mass of the star or planet and the distance to its centre in accordance with Newton's law of gravitation. Hence the gravitational field strength at distance r from the centre of a planet or star of mass M is GM / r^2. Thus, the gravitational field strength of a planet or star varies as the inverse of the square of the distance to the centre of the planet or star.

The gravitational field strength at the surface of a planet or star of radius R is GM / R^2. Hence the Moon's surface gravity is one-sixth of the Earth's surface gravity as the Earth's mass is 81 times greater and the Earth's radius is about 3.7 times that of the moon, making the Earth's surface gravity six times stronger ($= 81 / 3.7^2$). The surface gravity of a planet or a moon of a planet determines the escape speed from the planet.

see also...

Escape Speed; Newton's Law of Gravitation

Sun 1 – Structure

The sun is a glowing sphere of hot gases about 100 times larger than the Earth. The sun is a G class star and is known to emit energy due to radiation at a rate of 4×10^{26} watts. Its mass is known to be 2×10^{30} kilograms, its mean density is 1.4 times that of water and it has a well-defined light-emitting surface which is referred to as its **photosphere**. Its surface gravity is about 28 times that of the Earth's surface gravity. The temperature of the photosphere is known to be about 6000 K. Photographs taken using red filters show that the photosphere is covered by a layer of gases, no more than about 2000 km thick, referred to as the chromosphere. Eclipse photographs show that the sun is surrounded by tenuous gases that stretch far into space, referred to as the solar corona.

The interior of the sun contains a core in which nuclear fusion takes place and photons of gamma radiation are released in the process. These photons travel away from the core, interacting with fast-moving atomic nuclei and electrons until they reach the region referred to as the 'convective' zone where nuclei and electrons are combined in the form of atoms and ions. The outer boundary of this region forms the photosphere. The interior of the sun between its energy-producing core and the region where atoms and ions exist is referred to as the 'radiative zone'. The matter in this zone is a gas of dense unattached nuclei and electrons with too much kinetic energy to form atoms and ions. The inward pull of gravity on the matter in the radiative zone is counteracted by the outward pressure of this gas, provided the gas continues to be heated by a steady stream of radiation from the nuclear furnace at the core of the sun.

WARNING   Never look at the sun under any circumstances as it will damage your eyes permanently.

see also...

Nuclear Fusion; Stars 2

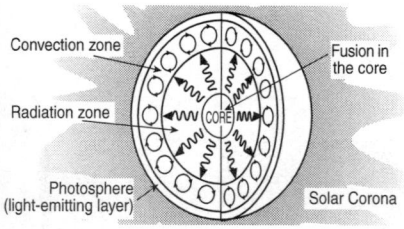

Convection zone

Radiation zone

Photosphere
(light-emitting layer)

Fusion in the core

CORE

Solar Corona

Inside the sun

Sun 2 – Solar Activity

High resolution photographs of the sun's photosphere show that it is granulated which means it is slightly blotchy. The blotches are convection cells about 1000 km wide. These convection cells are part of the convective zone which is thought to extend to a depth of about 0.3 × the solar radius. Each granule lasts for no more than a few minutes, as hot matter in each granule rises up its centre, cools off and falls down the sides. The chromosphere is a thin layer of gas much less dense than the photosphere and which covers the photosphere. The temperature of the chromosphere rises from about 4000 K at the top of the photosphere to more than 20,000 K at the top of the chromosphere. Jets of gas referred to as spicules rise and fall at high speed to heights of 10,000 km, disappearing after ten minutes or more. Spicules occur along the boundaries of supergranules which are irregular large groups of granules. Gas flows across the photosphere from the central regions of a supergranule to its boundaries where the gas falls back into the interior.

The solar corona is an envelope of hot gas surrounding the sun which extends by variable amounts in different directions from the sun. Its density is about a millionth of a millionth of the density of the photosphere and its temperature is of the order of millions of degrees. Ultraviolet images of the corona taken by the SOHO space probe have revealed coronal holes which act as vents for the solar wind. Massive eruptions from the corona throw vast amounts of gas into space at enormous speeds. In addition, enormous arches of solar matter, known as solar prominences, are thrown up from the photosphere, lasting days or even weeks before they die down. Solar flares are even more violent, emitting vast quantities of matter and radiation within a few minutes from regions of the photosphere which suddenly heat up to temperatures of more than 5 million K.

WARNING   Never look at the sun under any circumstances as it will damage your eyes permanently.

see also...

Solar Wind

Sun 3 – Sunspots

Sunspots are dark irregular spots and patches on the sun's photosphere, varying in size up to 10,000 km or more, lasting from hours to months before disappearing. Groups of sunspots are frequently created; each sunspot having its own dark centre at about 4000 K in temperature known as the umbra, which is surrounded by a less dark region at a temperature of about 5000 K known as the penumbra. Sunspots move across the solar disc because the sun is rotating steadily, taking about four weeks for each complete rotation. The further a sunspot is from the sun's equator, the longer it takes to go around the sun. Sunspots near the equator take about 25 days, whereas sunspots near the poles take about 35 days. The reason is that the sun is a ball of fluid and its rate of rotation decreases with latitude.

The number of sunspots on the photosphere increases from a minimum to a maximum and back to a minimum once every 11 years. Sunspot maxima occurred in 1989 and again in 2000 and will occur in 2011. Very few sunspots were evident in 1986. In each 11-year cycle, sunspots first appear 30 degrees north and south of the equator, gradually moving nearer the equator until after 11 years they appear on or near the equator before disappearing and reappearing at 30 degrees north and south of the equator. Sunspots are linked to solar magnetic fields as the sunspots on the trailing side of a group have reverse magnetic polarity in comparison to the sunspots on the leading side. Also, the magnetic polarity associated with the sunspots reverses once every 11 years when the magnetic polarity of the sun reverses.

Bright spots on the photosphere, referred to as plages, are observed just before sunspots appear. In addition, dark filaments seen near plages and sunspots are thought to be chromospheric material forced into arches by the effect of magnetic field. Where these arches are seen at the edge of the solar disc, they form prominences which can last for months.

see also...

Sun 1

Sundial

A horizontal sundial is designed to tell the time according to the position of the shadow of its gnomon on the dial when the sundial is aligned as shown below. The dial is graduated in hours as shown below with the midday position marked as 12.

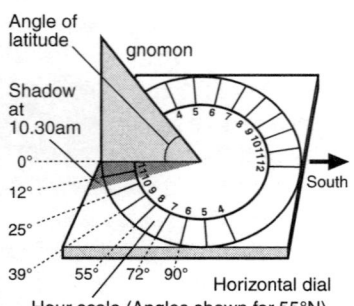

The sundial

The reading of a sundial differs from civil time which is the time in each time zone of the Earth. Civil time is determined by atomic clocks in designated scientific laboratories. A sundial reads local solar time (LST) which changes according to longitude. Zero longitude is defined as the meridian passing through Greenwich in England. Thus, a sundial at Greenwich gives Greenwich Solar Time (GST) which

differs from a sundial at longitude θ degrees by (θ/360) × 24 hours. For locations east of Greenwich, this difference is subtracted from LST to give GST; for locations west of Greenwich, the difference is added

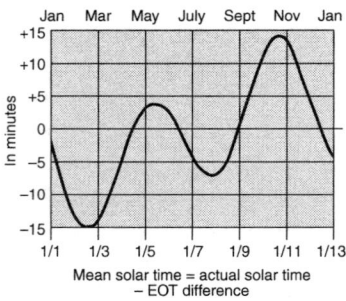

Mean solar time = actual solar time − EOT difference

The equation of time

on to give GST. Because the Earth's orbit is not exactly circular, the reading of a sundial can differ by as much as 15 minutes either way from its reading if the Earth's orbit was exactly circular. This difference, the Equation of Time (EoT), varies during the year as shown above. Greenwich Mean Time is GST − EoT.

see also...

Celestial Sphere 2; Sidereal and Solar Time

Supernova

A supernova is a star that ends its lifecycle in a massive explosion which can outshine an entire galaxy for months. The Crab Nebula M1 in Taurus is an irregular patch of glowing gas with filaments extending from the patch. The Crab Nebula is thought to be the result of a supernova explosion in 1054 at a distance of about 2000 parsecs from Earth. In addition to the Crab Nebula, only two other supernova have been detected in our home galaxy, the Milky Way. One of these, Tycho's star, occurred in Cassiopeia in 1572 and became as bright as Venus for over a year. The other one, Kepler's star, occurred in the constellation of Ophiuchus in 1604. An important supernova event was observed in 1987 in the Large Magellan Cloud which is an irregular companion galaxy of the Milky Way. This supernova, referred to as SN1987A, was a blue supergiant known as Sanduleak which suddenly exploded. The supernova brightened and reached magnitude 3 after a few months. Since then, it has gradually faded as the debris of the explosion continues to disperse at enormous speed.

Supernovae deficient in hydrogen are classed as Type I supernova as they have similar light decay curves. Other supernova including SN1987A were classed as Type II. Type I supernovae are further divided into three categories: a,b and c, according to the chemical elements detected. A Type Ia supernova in a distant galaxy is like a 'milestone' that can tell us how far away the galaxy is. This type of supernova occurs when a white dwarf attracts mass from a binary companion and suddenly collapses causing a shock wave to spread out through the outer layers which are blasted off in a cataclysmic explosion. The other types of supernovae are due to the collapse of stars of mass in excess of about eight solar masses, as these stars are unable to throw off excess mass above a certain limit. This was worked out in 1930 by Subrahmanyan Chandrasekhar who showed that a dying star would collapse if its mass was more than 1.4 solar masses. This became known as the 'Chandrasekhar limit'.

see also...

Evolution of Stars; Neutron Star; White Dwarf

Telescopes 1 – Refractors and Reflectors

A telescope is designed to make a distant object appear larger or to make a point object such as a star appear brighter. The simple refracting telescope consists of two convex lenses, the objective and the eyepiece. The objective forms a real image of a distant object in the focal plane of the objective. In normal adjustment, an observer looking through the eyepiece sees an enlarged virtual image of the real image.

that are too faint to be seen with a narrower objective. In addition, the wider the objective, the greater the amount of detail that can be seen in the image of an extended object. Large optical telescopes use a wide concave mirror as the objective to focus light. This is because large mirrors are easier to make and use than large lenses. A small mirror near the focal point of the concave mirror is used to reflect light into the eyepiece. Chromatic aberration is eliminated at the objective by using a concave mirror instead of a convex lens. In addition, if the concave mirror is parabolic in shape, spherical aberration is also eliminated.

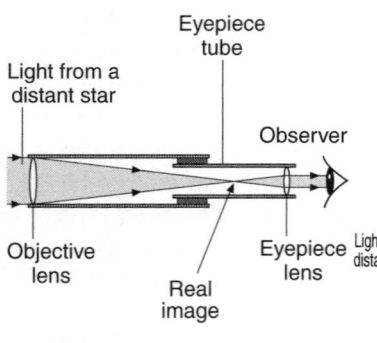

A refractor telescope

The wider the objective of a telescope, the more light that can be collected from a point object such as a star. Hence a wide objective telescope enables stars to be seen

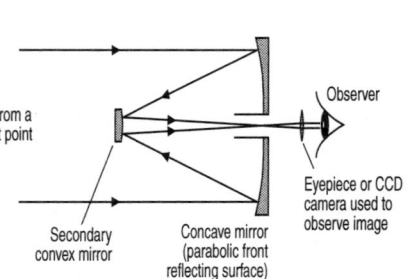

A reflector telescope

see also...
Eyepieces; Hubble Space Telescope

Telescopes 2 – Magnifying Power

The magnifying power is the factor which tells the user how many times larger the image is than the object. In normal adjustment, the magnifying power of a telescope $= \dfrac{f_0}{f_e}$, where f_0 is the focal length of the objective lens and f_e is the focal length of the eyepiece lens.

The magnifying power of a telescope should exceed one-eighth of the objective width in millimetres, otherwise not all the light entering the objective passes into the observer's eye, assuming the eye pupil of the observer is not greater than 8 mm.

A telescope collects more light than the unaided eye does because the objective lens is much wider than the eye pupil. This extra light-collecting power is important in astronomy. The light collected by the telescope depends on the area of the objective lens and the light collected by the unaided eye depends on the area of the eye pupil. Therefore, the light collected by the telescope increases in proportion to the square of the ratio of the diameters of the objective to the eye pupil. It can be shown that this diameter ratio is equal to the magnifying power M. Hence, using a telescope increases the amount of light from a point object by a factor M^2. For example, a 100 mm diameter telescope in comparison with a 10 mm eye pupil increases the light collected by a factor of 100 ($= 100^2/10^2$), corresponding to an extra 5 magnitudes. Such a telescope would therefore enable astronomers to see stars as faint as magnitude 11.

The width of the field of view of a telescope is usually expressed as the angle subtended to the eyepiece by the largest possible image that can just fit into the field of view. An extended object viewed through a telescope of magnifying power M appears M times wider and M^2 larger in area. Since the light collected increases by the same factor, the image brightness is the same in theory. In practice, absorption of light by the lenses reduces the light that enters the eye so the image could appear fainter using a telescope.

see also...

Magnification; Magnitude

Telescopes 3 – Resolving Power

The resolving power of a telescope is the angle between the lines of sight to two point objects such as two nearby stars that can just be distinguished. For example, if a telescope can just distinguish two stars at an angular separation of five seconds of arc, its resolving power is stated as five seconds of arc. Note that 3600 seconds of arc = one degree. The better the resolving power of a telescope, the greater the image detail that can be seen.

Diffraction of light at the objective lens causes the image of a point object to be smeared out. The Rayleigh criterion for resolving two nearby points, based on the theory of diffraction at a circular gap, states that two nearby point objects cannot be distinguished if their angular separation in arc seconds is less than $2.5 \times 10^5 \lambda/D$, where λ is the wavelength of light and D is the width of the objective lens. Thus a 100 mm diameter reflector telescope can resolve stars as close as one second of arc. In practice, ground-based telescopes larger than about 0.5 m in diameter do not achieve theoretical resolving power as the atmosphere also causes spreading of light by about 0.2 seconds of arc. The Hubble Space Telescope has an objective mirror diameter of 2.4 m and hence its theoretical resolving power is 0.04 seconds of arc. It provides far more detail than the same size telescope sited on the ground as it is unaffected by atmospheric refraction.

The resolving power of a radio telescope can be estimated from the Rayleigh criterion, given the wavelength and dish diameter. A 50 m dish operating at a wavelength of 0.1 m would be unable to resolve point objects any closer than about 0.1 degrees, much less than even a low-power optical telescope. By linking radio telescopes in different locations together, the resolving power of the linked telescopes can be made much greater than any of the individual telescopes.

> **see also...**
>
> *Hubble Space Telescope;*
> *Radio Telescopes*

Thermal Radiation

Thermal radiation is electromagnetic radiation emitted from the surface of an object due to its temperature. A glowing object emits light as well as infra-red radiation. Thermal radiation therefore includes light as well as infra-red radiation.

An object that absorbs all the electromagnetic radiation directed at it is referred to as a 'black body'. A star is a black body since any radiation directed at it is absorbed by it. The intensity of radiation from a black body was measured across a range of wavelengths for different temperatures, giving results indicated by the black body radiation curves shown below.

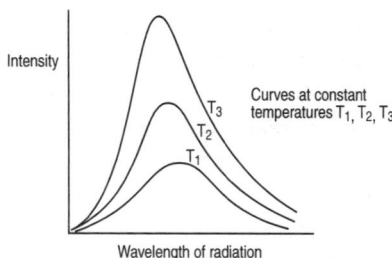

Black body radiation curves

Analysis of these curves led to the discovery of the laws of black body radiation which can be applied to stars to determine their surface temperatures and diameters.

Wien's Law states that the wavelength at peak intensity times the temperature is constant. Measurements show that this constant is 0.0029 metre kelvins. Thus, if the wavelength for the peak intensity of radiation from a star is measured, the surface temperature of the star can be determined. For example, the sun's spectrum is most intense at a wavelength of about 5×10^{-7} m giving a surface temperature of 5800 K.

Stefan's Law states that the total energy per second emitted = $\sigma A T^4$, where T is the surface temperature, A is the surface area and σ is the Stefan constant which equals 5.67×10^8 watts per square metre per Kelvin^4. Given the sun's surface temperature is 5800 K and its radius is 6.96×10^8 m, Stefan's Law gives its total energy emitted per second as 3.9×10^{26} watts.

see also...

Electromagnetic Radiation

Ultraviolet Astronomy and Beyond

Electromagnetic radiation stretches beyond the violet part of the visible spectrum into the ultraviolet region then into the X-ray and gamma ray regions. The Earth's atmosphere absorbs these forms of electromagnetic radiation beyond the visible spectrum so sources of such forms of radiation can only be detected using suitable instruments on board satellites in orbit above the Earth's atmosphere.

Ultraviolet astronomy was first carried out in detail from the International Ultraviolet Explorer between 1978 and 1996. The Extra Ultraviolet Explorer was launched in 1992 to observe sources at much shorter wavelengths from about 10 nm to 70 nm. The Hubble Space Telescope is also able to observe ultraviolet radiation sources but not at wavelengths shorter than about 120 nm.

X-ray sources include possible black holes and exploding stars as well as hot gases in space. The first survey of X-ray sources in the sky was obtained in 1971 using the Uhuru satellite which produced the first evidence leading to the discovery of X-ray binary pulsars. Further discoveries from satellite-based X-ray telescopes include X-ray bursters which produce bursts of X-rays at relatively long intervals. An X-ray telescope works by reflecting X-rays off highly polished metal plates at 'grazing incidence' onto a suitable detector.

Gamma ray bursters were discovered over 30 years ago when defense satellites being used to pinpoint nuclear weapons tests, discovered gamma ray bursts from space in random directions. The Compton Gamma Ray Observatory was launched from the Space Shuttle Atlantis in 1991 to study gamma sources in space. In 1997, the much more precise X-ray detector on satellite BeppoSAX, pinpointed a gamma ray burst which was then located using optical telescopes. The red shift of this source was measured and its distance was found to be billions of years away. Further discoveries have since pinpointed gamma ray bursters up to ten billion light years away.

Uranus

Uranus was discovered by William Herschel in 1781, although it had been marked on star maps before then but was thought to be no more than a dim star. Herschel plotted its position relative to other stars and concluded that it must be a planet beyond Saturn because it was moving very slowly at a rate of about four degrees per year through the constellations. In fact, Uranus orbits the sun at a mean distance of 19 AU taking 84 years for each complete orbit. Its diameter is four times that of Earth and its mean density is about 1.3 times that of water. Its surface gravity is 0.9 times the Earth's surface gravity and its surface temperature is 55 K.

Observations from Earth show it as a featureless greenish-blue disc. Voyager 2 flew past Uranus in 1986 and confirmed its featureless surface and its atmospheric composition which was found to be about six parts hydrogen to one part helium, with small amounts of heavy elements and methane which causes its bluish colour.

Herschel also observed two moons orbiting Uranus in a plane at right angles to the orbit of Uranus around the sun, indicating that Uranus has a spin axis that is tilted at 90 degrees to its orbital axis. In fact, the planet is known to have 15 moons, all orbiting in the same plane. Voyager 2 confirmed its axial tilt which has since been found to be 98 degrees. Because the axis of Uranus always points in the same direction, each Uranian pole experiences decades of continuous darkness followed by decades of continuous daylight. The tilt of Uranus's axis is thought to have been caused by a collision with a massive object. A similar collision may account for the fragmentary appearance of the surface of Miranda, one of the moons of Uranus.

Voyager 2 also confirmed the existence of a faint Uranian ring system which had first been detected a decade earlier when astronomers observed Uranus as it occulted (i.e. passed in front of) a star. The star 'blinked' unexpectedly just before and just after occultation, indicating the presence of rings round Uranus which blocked the star light when they passed in front of the star.

Variable Stars

variable star is a star with variable brightness. An eclipsing binary is a variable star because its brightness drops temporarily each time one of its components eclipses the other one.

Variable stars that vary in brightness without the periodic dips characteristic of eclipsing binaries are 'intrinsic variables' as the brightness changes are due to internal changes. For example, the brightness of Mira in the constellation Cetus changes smoothly from second magnitude (and hence easily visible) to tenth magnitude and back once every 331 days.

Cepheid variables range in periods from about one day to over 100 days, changing brightness by not much more than 1 magnitude. A cepheid variable brightens faster than it fades. Cepheids are known to be pulsating stars as their spectral lines shift back and forth repeatedly. The period of a cepheid variable depends on its mean absolute magnitude which is why cepheid variables can be used as distance markers.

RR Lyrae stars vary in brightness similar to cepheid variables and they are also thought to be pulsating stars. However, they have periods of the order of hours rather than days and they are class A to F stars, whereas cepheid variables range from class G to M. RR Lyra are found mostly in globular clusters.

RV Tauri stars have periods from 30 to 150 days. These stars brighten smoothly and fade out unevenly. T tauri stars vary irregularly in brightness by several magnitudes. They are found only in dust and gas clouds which probably means they are new stars.

Long-period variables like Mira vary in brightness with periods from 100 to 1000 days. The change of magnitude can be 10 or more magnitudes.

Nova are stars that brighten by many magnitudes in a very short time, fading slowly and usually returning to their former brightness.

see also...

Cepheid Variables; Clusters of Stars; Magnitude; Stars 2

Venus 1 – Orbital Features

Venus at its brightest is brighter than any other object in the sky except the sun and the moon. Its brightness is due partly because it is permanently covered in white cloud so it reflects sunlight very effectively. Venus orbits the sun at an average distance of 0.72 AU in an orbit that is almost circular. It takes 225 days for each orbit and spins in a retrograde direction at a steady rate of 243 days per turn. Its diameter is 0.95 times that of the Earth and its mean density is 5.2 times that of water.

Venus as seen from the Earth is never further from the sun than 47 degrees, its maximum elongation. Thus, Venus can never be seen much more than three hours before sunrise or three hours after sunset. When seen before sunrise, it is at 'western elongation' since it is west of the sun; when seen after sunset, it is at 'eastern elongation' since it is east of the sun. The brightness of Venus varies because its distance from the Earth can change from as little as 0.28 AU when it is at 'inferior conjunction' between the Earth and the sun to 1.72 AU when it is at superior conjunction on the opposite side of

the sun to the Earth. As it moves from superior to inferior conjuction and onto the next superior conjunction, its appearance passes through a cycle of phases from 'full' to 'crescent' and back to 'full'. However, the angular width of its disc also changes due to its distance from Earth changing so its appearance changes from a small 'full Venus' disc to a large 'crescent Venus' and back. Its maximum brightness is when it is near maximum elongation because a high proportion of its sunlit surface is then visible and it is not too far away.

The orbit of Venus is inclined at about 3 degrees to the Earth's orbit. When Venus is at inferior conjunction, it is usually above or below the solar disc because of the inclination of its orbit. However, on rare occasions referred to as 'transits', it may be seen moving across the solar disc as a black dot. The last transits were in 1874 and 1882, the next two will be in 2004 and 2012.

see also...

Planet; Planetary Orbits

Venus 2 – Surface Features

Venus is permanently and completely covered by cloud. The rotation period of Venus was determined first using radar pulses from Earth directed at Venus. These pulses are partly reflected by the surface of Venus (as radar penetrates the clouds) and so can be detected at the Earth. Because Venus is rotating, the wavelength of the pulses reflected from Venus are shifted to longer wavelengths by the receding half of Venus and shifted to shorter wavelengths by the advancing half of Venus. By measuring the shift of wavelength of the reflected pulses, the rate of rotation of the planet was shown to be retrograde with a period of 243 days.

The surface gravity of Venus is 0.9 times that of Earth and its surface temperature is about 750 K. Its atmosphere is 96 per cent carbon dioxide, 4 per cent nitrogen and smaller quantities of sulphur dioxide, hydrogen sulphide and other chemicals. Measurements made by spaceprobes directed into the atmosphere of Venus have shown the pressure is 90 times that of the Earth's atmosphere. The Venusian clouds, mostly sulphuric acid, are about 20 km thick at a height of about 50 km above the surface, leaving a clear atmosphere from the surface to a height of about 30 km. The surface temperature is much higher than expected for an object at 0.72 AU from the sun; this is because its cloud layer traps heat from the sun, thus causing a 'greenhouse effect' in its atmosphere.

The surface of Venus was mapped out in detail between 1990 and 1992 by Magellan, an orbiting spacecraft fitted with a radar system designed for the purpose. The images from Magellan revealed solidified lava flow channels from volcanoes which may not be extinct and which are likely to be responsible for the sulphur content of the atmosphere.

see also...

Atmosphere (Earth's); Radar Astronomy; Strength of Gravity

White Dwarfs

White dwarf stars are much less luminous than the Sun but they are much hotter than the sun because they are white hot. According to Stefan's law of radiation, the hotter a star is, the greater is the energy per second per unit surface area emitted by it. Therefore, a white dwarf star emits much more energy per second per unit surface area than the sun but, because the total energy per second emitted by a white dwarf is much less than by the sun, the surface area and hence the diameter of a white dwarf must be much less than that of the Sun. The term 'dwarf' is used to mean that the star is much smaller in diameter than the sun. Sirius B, the binary companion of Sirius, was one of the first white dwarf stars to be discovered.

A white dwarf star is formed when nuclear fusion ceases inside a low-mass giant star. The giant star may eject a considerable part of its outer layers as so-called planetary nebulae, leaving a hot dense core as the white dwarf which gradually fades out over billions of years. The radius of a white dwarf decreases as its mass increases upto the Chandrasekhar limit of 1.4 solar masses. A star core of mass greater than 1.4 solar masses does not become a white dwarf as it is unstable and explodes as a supernova. A white dwarf star that is accompanied by a companion star in a binary system attracts mass from its companion and explodes as a nova or a supernova when its mass reaches the Chandrasekhar limit. Nuclear fusion starts again when the Chandrasekhar limit is reached, causing the outer layers to be thrown off in the case of a nova or causing the entire star to explode in the case of a supernova. However, an isolated white dwarf is stable and radiates its internal energy, eventually becoming a cold invisible dwarf star. With a considerable amount of carbon in such a star, some astronomers reckon such a star might be like a gigantic diamond ball!

see also...

Evolution of Stars; Nuclear Fusion; Supernova

Glossary

Absolute magnitude M the apparent magnitude of a star if it was 10 parsecs (= 300 million million km) away

Apparent magnitude M the brightness of a star on a scale from m = 1 for the brightest to m = 6 for stars just visible on a clear night. Every difference of 5 magnitudes corresponds to a change of light energy by a factor of 100

Arc second an angle equal to one-sixtieth of one-sixtieth of one degree

Astronomical unit (AU) the mean distance from the sun to the Earth, equal to 149.6 million km.

Conjunction the position of a planet on its orbit when it is the same direction as the sun

Culmination when a star or planet reaches its maximum altitude above the horizon

Declination the angle from a star to the nearest point on the Celestial Equator

Doppler effect change of wavelength of light from a source due to its motion towards or away from the observer

Ecliptic the sun's annual path around the Celestial Sphere

Elongation the angle between the lines of sight to an inner planet and to the sun

First point of Aries the point on the Celestial Equator which the sun crosses in spring (around 21 March)

Main sequence the main band of stars on the Hertzsprung Russell diagram. A main sequence star produces energy as a result of converting hydrogen to helium by nuclear fusion in its core

Meridian the great circle on the Celestial Sphere which passes through both poles and directly overhead

Opposition the position of an outer planet on its orbit as seen from Earth when it is in the opposite direction to the sun

Parsec (pc) 1 parsec = 3.26 light years = 30 million million km. The parsec is defined as the distance to a star which subtends an angle of 1 arc second to the Earth and sun

Photon a wave packet of electromagnetic radiation

Population I and II stars population I stars are hot, blue metal-rich young stars; population II stars are old metal-deficient cooler red stars

Right ascension the angle eastwards along the Celestial Equator from the First Point of Aries to the great circle through a star

Further reading

Abrams, Bernard and Moore Patrick *Astronomy: Selected Topics* (Stanley Thornes, 1989)

Breithaupt, Jim *Einstein: A Beginner's Guide* (Hodder & Stoughton, 2000)

Breithaupt, Jim *Teach Yourself Cosmology* (Hodder & Stoughton, 1999)

Moore, Patrick *Teach Yourself Astronomy* (Hodder & Stoughton, 1999)

Also available in the series